DEPARTMENT OF THE ENVIRONMENT

Assessment of Groundwater Quality in England and Wales

Prepared for Department of Environment by
Sir William Halcrow and Partners Ltd.
in association with
Laurence Gould Consultants Ltd.
July 1988

LONDON: HER MAJESTY'S STATIONERY OFFICE.

© *Crown copyright 1988*
First published 1988

ISBN 0 11 752169 8 ✓

DEPARTMENT OF THE ENVIRONMENT – RESEARCH CONTRACT PECD 7/7/227

ASSESSMENT OF GROUNDWATER QUALITY IN ENGLAND AND WALES

FINAL REPORT

CONTENTS

ACKNOWLEDGEMENTS

The Consultants would like to express their appreciation for the assistance and cooperation provided by the many representatives of the British Water Industry consulted during this study. Our thanks are extended to the following organisations:

The Department of the Environment

The Water Authorities Association

The Regional Water Authorities

The Mid Kent and Colne Valley Water Companies

The Ministry of Agriculture, Fisheries and Food

The Welsh Office

The Natural Environment Research Council

The Water Research Centre

British Geological Survey

Property Services Agency

GLOSSARY

The purpose of the glossary of terms given below is to provide an explanation for abbreviations and technical expressions used in the report. It is recognised that some terms may be used in a wider context and the explanations given below are intended to represent the context in which the words are used herein.

Attenuation	:	the weakening of a pollution plume by the processes of dilution, dispersion, adsorption etc.
Baseflow	:	the contribution to surface water flow from groundwater.
BGS	:	British Geological Survey.
CEGB	:	Central Electricity Generating Board.
Confined Aquifer	:	an aquifer in which groundwater is confined under pressure by overlying and underlying strata.
Connate Groundwater	:	groundwater which is of the same age as the strata in which it occurs.
Contamination	:	of groundwater; the introduction of a potentially polluting substance which measurably increases the existing concentration of that substance in groundwater.
Cross Contamination	:	Sampling error caused by introduction of substances from extraneous sources into the sample.
DHSS	:	Department of Health and Social Security.
Diffuse Contamination/ Pollution	:	contamination/pollution of groundwater over a wide area eg fertiliser application on arable land.

Drift	:	Superficial deposits of Quaternary and Recent ages.
EC	:	European Community.
Extensification	:	reduced intensity of a farming system either by altering cropping and stocking or by reducing the level of inputs.
Facies	:	the character of a sedimentary deposit in terms of its composition, texture etc, (eg sandy facies); a term often used to distinguish one deposit from another of the same age.
GCMS	:	gas chromatography mass spectrometer
Halite	:	rock salt.
Hydrogeochemistry	:	the study of the chemistry of soil/rock/water systems.
Hydrochemical	:	pertaining to the chemical composition of water.
Iron Species	:	iron compounds in different valency states, for example ferric (trivalent) and ferrous (divalent) compounds.
Karstic	:	of limestone, characterised by fissures, caves and underground channels.
MAFF	:	Ministry of Agriculture, Fisheries and Food.
mg/l	:	units of concentration, milligrammes per litre.
Ml/d	:	units of flow, megalitres per day.
Multi-Point Contamination/Pollution	:	A series of point-sources producing the same kind of pollution.

Multi-Port Sampling	:	sampling of groundwater at various depths in a borehole from discrete portals opposite a specific horizon.
NERC	:	Natural Environment Research Council.
NIREX	:	Nuclear Industry Radioactive Waste Executive.
Observation Borehole	:	a borehole (usually small diameter) used for the monitoring of groundwater levels and groundwater quality.
Open-Hole	:	a borehole containing a column of water open to the atmosphere.
Outcrop	:	pertaining to a rock formation, where it is exposed at ground surface or occurs immediately below the soil horizon.
Pesticides	:	a family of substances, used for destroying pests, including insecticides, herbicides, fungicides etc.
Piezometer	:	a means, usually a borehole, by which groundwater heads are measured.
Point Contamination/ Pollution	:	contamination/pollution of groundwater from a single source, eg a landfill site.
Pollution	:	of groundwater, to make unhealthily impure by the introduction of substances.
Private Water Supplies	:	those water supplies to users which are not operated by the public water undertakings (the regional water authorities and the water companies).

Public Water Supplies	:	those water supplies to users which are operated by the public water undertakings.
Raw Water	:	water which is in its natural chemical state and has not been treated in any way.
Recharge Zone	:	an area over which an aquifer's groundwater resources are replenished, commonly by infiltration of rainfall at aquifer outcrop.
Redox	:	pertaining to oxidation - reduction type chemical reactions.
River Regulation/ Augmentation with Groundwater	:	schemes by which groundwater is artificially discharged to rivers normally to increase summer flows and to maintain surface water abstraction sources.
Saturated Zone	:	of an unconfined aquifer, that zone below the water table where the pore spaces in the soil/rock matrix are saturated with water.
Set-Aside	:	taking a proportion of farm land used for growing commodities in surplus out of production, usually leaving it fallow.
Triazines	:	Nitro-organic compounds used in weedkillers.
Trihalomethanes	:	simple halogenated organic compounds frequently associated with solvents.
Unconfined Aquifer	:	an aquifer in which groundwater possesses a free surface (water table) open to the atmosphere.

Unsaturated Zone	:	of an unconfined aquifer, that zone above the water table where the pore spaces in the soil/rock matrix are unsaturated with respect to water.
WHO	:	World Health Organisation.
Well Head Determination	:	measurement of chemical properties of groundwater as it is pumped out of a borehole at the well site in an attempt to measure parameters prior to the effects of degassing etc. e.g dissolved oxygen and carbon dioxide, alkalinity, pH etc.
Winterbourne	:	the section of a groundwater-fed stream or river which ceases to flow at certain times of the year due to the seasonal recession of the water table.
WRC	:	Water Research Centre.

SUMMARY

This research contract was commissioned by the Department of the Environment in early 1987, and substantially completed in the period June-December of that year. The principal objective of the study was to provide an overview of the present state of groundwater quality in England and Wales, with the emphasis on groundwater for use as drinking water. Additional requirements were to highlight existing and potential problems for managing and protecting groundwater, and to identify issues for further examination, including research.

The findings of the study are based on a review of relevant literature together with information and views obtained from interviews with representatives of organisations with a professional interest in groundwater quality. Those interviewed included representatives of each of the 10 regional water authorities, two water companies, government departments and research institutions. Because of the very large amount of published information available it was necessary to be selective with respect to documents used for detailed study.

An assessment was made of the importance of groundwater as a source of drinking water. At present, as shown in Figure S1, about 31% of drinking water in England and Wales comes from groundwater, an amount which has shown little change for the last 10-15 years. However, there is considerable variation between the regional water authorities in utilisation of groundwater as shown in Figure S.2, from 74% (Southern) to 10% or less (Northumbrian and Welsh), while the Anglian, Severn-Trent, Thames and Wessex Water Authorities are each 40%-50% reliant on groundwater for public supply. Also, even within the authorities which are the lowest

i

TOTAL WATER SUPPLY IN ENGLAND AND WALES (LICENSED) IN 1985

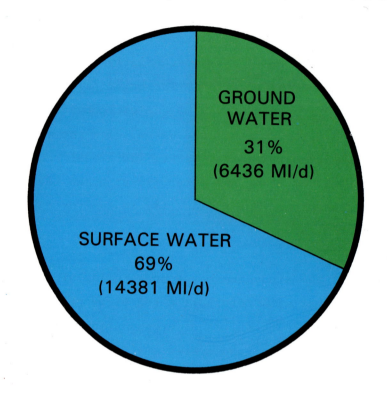

GROUNDWATER USAGE (LICENSED) IN ENGLAND AND WALES IN 1985

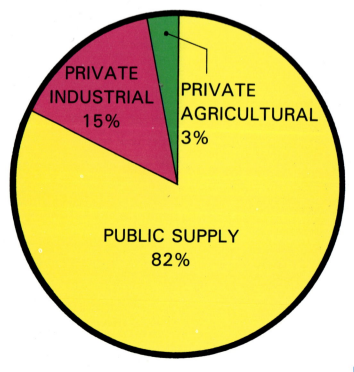

Halcrow

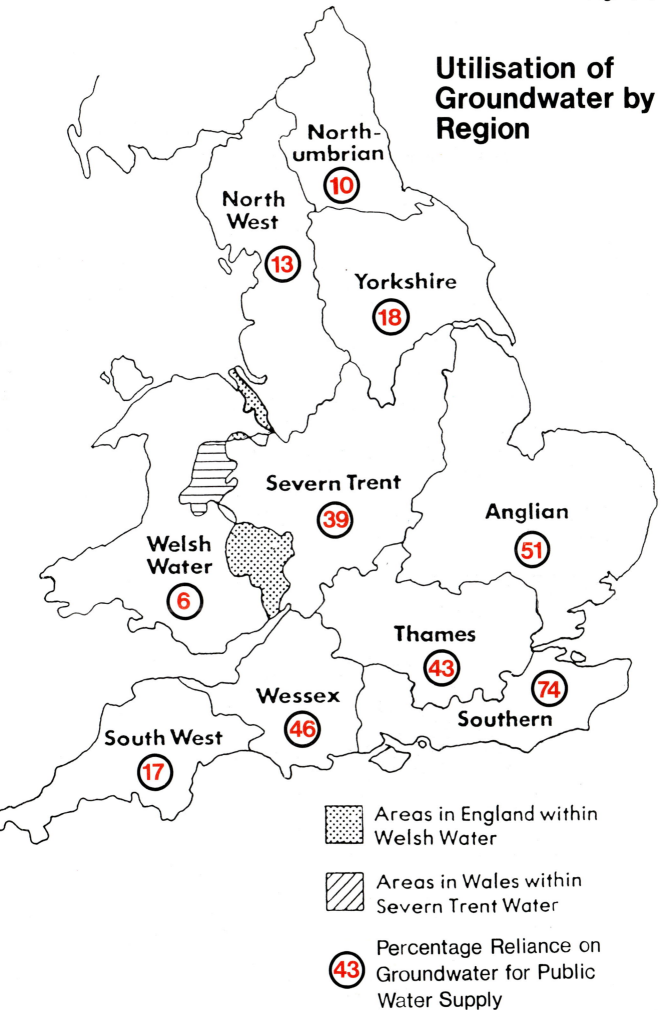

Utilisation of Groundwater by Region

North-umbrian (10)

North West (13)

Yorkshire (18)

Severn Trent (39)

Anglian (51)

Welsh Water (6)

Thames (43)

Wessex (46)

Southern (74)

South West (17)

Areas in England within Welsh Water

Areas in Wales within Severn Trent Water

(43) Percentage Reliance on Groundwater for Public Water Supply

groundwater users, there are towns which are totally reliant on groundwater.

Available information clearly demonstrates the national, regional and strategic local importance of groundwater as a source of drinking water. The cost of replacement of these sources should they become unusable due to pollution would be substantial.

Monitoring of groundwater quality in England and Wales is mainly carried out by the water authorities as part of routine operational activities. Under these circumstances the special needs of groundwater sampling and analysis are not always met. Additional data have been provided by groundwater resource investigations and management, pollution investigations and research projects. During the last 15 years a better understanding of the hydrogeochemical conditions in the major British aquifers has been established. This understanding includes controls and processes affecting the distribution of naturally present substances as well as contaminants.

There is no requirement under British legislation to monitor the quality of groundwater at source, but routine monitoring is usually carried out to the requirements of the EC Directive 80/778/EC on the Quality of Water Intended for Human Consumption. Monitoring for some organic compounds has created problems for the water undertakers because of the costs of sampling and analysing for complex substances whose maximum admissable concentrations (MAC) are significantly less than one part per billion. For groundwater monitoring to be of value, the sampling, analysis and interpretation should take account of the capacity of the groundwater to change chemically when withdrawn from the aquifer, when it may no longer represent in-situ aquifer conditions.

A series of activities can be identified which are a threat to groundwater quality. These are principally agricultural and industrial activities, waste disposal and accidents. The time-related aspect of these activities and their consequences in terms of groundwater quality is not always easy to determine. Some current groundwater contamination problems relate to events which occurred many years ago.

Agricultural activities are a source of contamination over wide areas of England and Wales. Contamination from diffuse sources is the most common problem, the main substances being nitrate and, possibly, pesticides of various kinds. Nitrate in groundwater originates mainly from fertiliser application and has been the subject of much research in recent years. The present situation is that the levels of nitrate in groundwater are continuing to rise in many areas, to approaching or beyond MAC levels. Some potable supply sources have already been lost and others are usable only by blending with lower nitrate waters. Because of the wide areas affected and the high cost of denitrification treatment, the nitrate problem is of serious concern.

The pesticide problem is more complicated. The MAC level of pesticides is 0.1µg/1 (one part in ten billion) of any single pesticide, and no more than 0.5µg/1 of pesticides in total. These levels are close to current levels of detection which raises particular sampling and analysis difficulties. In addition, the cause and effect relationship between agricultural use of pesticides and their occurrence in groundwater is not straightforward. For technical and economic reasons farmers tend to adhere closely to recommended application rates for pesticides while use for non-agricultural purposes is less well controlled and documented. Also, the national situation on the extent of pesticides in groundwater is not yet fully understood. However,

although the relative importance of agricultural and
non-agricultural sources of pesticide contamination has not yet
been firmly established, their increasingly common occurrence in
groundwater, in conjunction with the low MAC levels, is also cause
for serious concern.

Landfill waste disposal sites are a commonly occurring potential
source of groundwater pollution through the production of
leachate. A wide variety of contaminants may be introduced
depending upon the type of waste tipped. In most water
authorities, landfill is regarded as the traditional threat to
groundwater quality, though there are few examples of sources
being lost as a result of proximity to landfill sites.

The siting of landfills is controlled through legislation under
which the water authorities are consulted. In addition, codes of
practice have been introduced by the Department of the Environment
concerning the design and methods of working of landfills.
Nevertheless, the validity of important concepts in landfill
technology such as 'dilute and disperse' are being seriously
questioned by groundwater resources specialists, both for
technical reasons and in view of recent EC Directives.

Many industrial activities involve the transport, handling,
storage and disposal of a wide variety of substances which have
the potential to cause groundwater pollution. Transport of
materials creates the risk of pollution through accidents and
spillages on roads, many miles of which cross the major aquifers
and are drained by soakaways. There are also several examples of
pollution caused by pipeline fractures and leaks at the base of
chemical storage tanks. Careless or inadvertent disposal of
materials may also result in groundwater pollution. An example is
the widespread occurrence in the Permo-Triassic sandstone aquifers

of the West Midlands of organic contaminants thought to originate
from solvents and degreasing agents used in car manufacturing in
that area.

Many of the contaminants resulting from the industrial activities
described above are organic compounds. Contaminants of this type
frequently have MAC levels less than 0.1µg/l such that a large
slug of the substance in concentrated form may, depending upon its
solubility, be capable of polluting groundwater over a wide area.
Also, hydrogeological systems exist where the transmission of the
pollutants to a source of abstraction would be extremely rapid,
with little prospect of attenuation of the pollution plume. In
general, the increasing discovery of organic compounds of
anthropogenic origin in groundwater is cause for concern,
particularly in view of the current MAC levels.

Other sources of groundwater contamination have also been
identified. Coal mine drainage can cause groundwater pollution and
cases have been reported. The main threat arises through surface
water - groundwater interaction. Point source pollution as a
result of agricultural activities may cause local pollution
problems, mainly through careless disposal of effluents and
chemicals. Acid rain may cause problems in aquifers with low
buffering capacity while surface water - groundwater interaction
is a potential threat where the river is polluted and there is a
hydraulic gradient towards the aquifer. Saline intrusion is
considered to be under control in terms of coastal ingress of salt
water and old saline groundwaters in inland aquifers. Pollution of
small local aquifers from septic tanks may also be an issue.

Nevertheless, since large quantities of groundwater continue to be
abstracted, and supplied to the consumer at an acceptable standard
of quality following minimal treatment, it is concluded on present

evidence that the current state of groundwater quality in England and Wales is good. However, serious problems have been identified which threaten this situation, particularly nitrate and organic compounds. Additional investigations may show the extent of contamination by organic compounds to be more widespread than is known at present. It is likely therefore that use of groundwater sources for potable supply will increase substantially in cost, ceasing to be the "cheap option" which it has been for many years.

Existing legislation concerning groundwater quality includes Acts of Parliament and EC Directives together with less formal guidelines and codes of practice. Under the 1973 Water Act the regional water authorities are charged with the duty of providing a wholesome supply. Where appropriate this responsibility is discharged to a statutory water company. The present reference standard for drinking water quality is the EC Drinking Water Quality Directive 80/778/EEC relating to the Quality of Water Intended for Human Consumption.

Legislation concerning groundwater pollution is more complex. Under the Control of Pollution Act 1974, Part II, (COPA II) a water authority can specify underground waters for protection against discharges which may give rise to pollution. More recently the EC Directive 80/68/EC on Protection of Groundwater Against Certain Dangerous Substances has been issued, which sets out to control the discharge of specified (listed) substances. The UK is implementing this EC Directive through COPA II, though there is an inconsistency between the two. Whereas the Directive protects all groundwater, COPA II refers only to groundwaters specifed by the water authority. As a result the authorities are specifying larger areas for protection of groundwater against pollution than may have been envisaged under the Act.

COPA 1974 Part I deals with the disposal of waste to land, in recognition of its potential for pollution. Consultation procedures for the issue of waste disposal licences are established which involve the regional water authorities. Reinstatement and aftercare of the site is controlled under the 1971 Town and County Planning Act, although under this Act the water authorities are not statutory consultees. The requirement for strengthening of the legislation in this area has been recognised.

The use of sewage sludge in agriculture is a possible source of groundwater pollution which has attracted interest, although there appears to be little case-history evidence of pollution actually occurring. Nevertheless the potential threat has been recognised and the Department of the Environment together with the water authorities are currently preparing a code of practice to cover this activity.

Aquifer protection in England and Wales is affected by the statutory requirements reviewed above, together with less formal policies. The statutory requirements lead to the designation of aquifer protection zones, which usually relate only to areas close to wells. However, in some water authorities an Aquifer Protection Policy (APP) has been formulated which covers a whole catchment or aquifer. APP's currently in operation take into account factors such as vulnerability of groundwater in a particular area to pollution and likely travel times of pollutants to wells. Through the operation of a comprehensive APP, highest risk activities can be located in the lowest vulnerability areas. Also, planning and waste disposal authorities are made aware of the water authorities view at an early stage, which renders the decision making process more efficient.

Privatisation of the water authorities, which is anticipated within the term of the present parliament, will involve the creation of a National Rivers Authority (NRA). The responsibilities of the NRA will include conservation and prevention of pollution in groundwater, through specification of protection zones. Some functions, such as sampling and monitoring of groundwater, may be carried out privately under contract.

Research into groundwater quality is principally funded by the Department of the Environment, the Natural Environment Research Council and the water authorities. The main organisations undertaking the work are the Water Research Centre, British Geological Survey and the authorities themselves. Wide ranging consultative groups are of value in coordinating multi-disciplinary research in groundwater quality.

Recent research into inorganic substances in groundwater has provided a better understanding of natural hydrochemical conditions and numerical modelling has contributed. The research has included studies of salinity distributions and acid rain. Trace element and isotope studies have also been useful.

Nitrate in particular has been the major groundwater quality research topic in the last 10 years and much has been achieved, to the extent that modelling of nitrate transport allows site specific forecasting. Research is continuing into possible microbiological effects on nitrate movement, relationships between nitrate concentrations and land use, and post-abstraction and within-aquifer denitrification.

Organic quality research has only recently commenced in any serious form. Investigations are underway into sampling and analysis for the main organic contaminants identified. Landfill research has been extensive but mainly directed towards domestic landfills operating under natural ground attenuation principles.

A series of research requirements can be clearly identified, and past UK achievements have been such that a foundation exists on which future programmes can be built.

The principal conclusions of this report are as follows:

o groundwater provides approximately 30% of drinking water in England and Wales and is a major source of water both nationally and strategically;

o groundwater quality is generally good but problems associated with contamination are increasing and the value of the resource as a considerably less expensive alternative to surface water is seriously threatened;

o the occurrence, distribution and behaviour of many important contaminants is not well understood and improvements in monitoring practice are required;

o the most important contamination problems are associated with nitrate, pesticides and various organic solvents;

o legislation for the protection of groundwater appears to have been effective but shortcomings can be identified; a national Aquifer Protection Policy would be beneficial;

o recent and current research into groundwater quality has made a signficant contribution in many areas, and provides an excellent foundation of knowledge and experience which can be built upon;

The priority recommendations are as follows:

o   management of the nation's groundwater resources should be
     coordinated more effectively;

o   a national aquifer protection policy and the monitoring of
     groundwater quality on a national network should be
     established, based on hydrogeological principles;

o   the water authorities should become statutory consultees
     under the planning aspects of COPA I with respect to
     aftercare of sites, and other planning matters which have a
     bearing on water quality;

o   due recognition should be given to the views within the
     water supply industry that the levels of some constituents
     specified in the EC Drinking Water Quality Directive are
     unrealistic in terms of their perceived effect on human
     health.

o   research on the following topics should be instigated:

     (a) development of effective groundwater sampling systems

     (b) scientific methodology for landfill leachate and
          borehole investigations

     (c) pollution problems in shallow aquifers

     (d) hydrochemical processes controlling migration of
          organic pollutants

(e)   baseline data relevant to contaminants in groundwater

(f)   the effect of land use change and agricultural practice
      on nitrate and pesticides in groundwater

1. INTRODUCTION

1.1 Background

This study was carried out in the form of a Research Contract as part of the water research programme commissioned by the Department of the Environment.

In November 1986 consulting firms and similar organisations were requested to express their interest in carrying out the study, based on draft terms of reference on which they were invited to comment. In February 1987, prequalified consultants were invited to submit costed proposals to undertake the work, in accordance with terms of reference which by that stage had been finalised. These terms of reference are included in Appendix A.

Three consulting firms were subsequently short-listed, and inverviewed in May 1987 by representatives of the Department. In June, Sir William Halcrow and Partners Ltd were advised of their appointment as Research Contractor, with a contract duration of 6 months extending from 1st July to 31st December 1987. The contract was subsequently extended to 31st May 1988.

1.2 Conduct of the Study

1.2.1 Organisation

The project was implemented by Halcrow from its offices at Burderop Park, Swindon, Wiltshire. In execution of the project Halcrow worked in association with Laurence Gould Consultants Ltd of Warwick, and Prof J W Lloyd of the Hydrogeology Section, Department of Geological Sciences, University of Birmingham.

A staff list and organogram is given in Appendix B.

1

1.2.2   Project Execution

Work by the consultants commenced on 23rd July 1987, with a meeting of members of the project team to review the objectives of the project and issues to be considered, to confirm the programme of work, and to consider names of organisations and individuals to be consulted.

The list of organisations was integrated with an equivalent list provided by the Department and formed the basis of a submission dated 5th August comprising:

o   a proposed programme of work;
o   an indication of which team members proposed to meet the various individuals/organisations concerned;
o   additional organisations to be consulted;
o   information to be obtained;

The submission was approved by the Department and meetings were arranged during August to November 1987.

An Inception Report outlining progress to date and preliminary findings was issued to the Department on 4th November.

Further review of the large amount of data and reports obtained was carried out during November/December as part of preparation of the Final Report.

A draft Final Report was submitted to the Department for review in December 1987, and the consultants reviewed their findings at a presentation given in Romney House on 11th January 1988. The amended Final Report was submitted in July 1988, completing the requirements of the research contractor for the study.

1.3   Sources of Information

The two main sources of information on which the findings of the study are based were personal interviews with individuals representing organisations with an interest in groundwater quality issues, together with relevant reports and documents.

Organisations were approached by the Department and requested to nominate an individual with whom members of the project team were to make contact to discuss the study. Details of meetings held are given in Figure 1.1. These meetings provided a considerable amount of information and views from people directly involved in groundwater quality management and research.

The amount of information currently available relevant to the issue of groundwater quality in England and Wales is immense. The necessity to be selective with regard to choice of documents for detailed critical review was therefore recognised from the earliest stages of the project. Documents containing a summary or review of a particular subject or issue were therefore particularly useful.

A selected list of references is provided at the end of this report.

Figure 1·1

PRINCIPAL ORGANISATIONS CONSULTED

Organisation	Meeting Held with:	Sphere of Interest	Visited by:	Other Repreesentatives Consulted / Other Attendees
DOE WATER DIRECTORATE				
Water Technical Division	M.G.Healey A.Goodman C.E.Wright	Head of Division ; quality standards, research policy, sludge disposal;	EC/DWMJ EC/DWMJ/CHM,JWL EC/CHM	DWC Rodda
Water Quality Division	R.Otter	Protection of resources,standards	EC/DWMJ,JWL	
OTHER DOE DIRECTORATES,DIVISIONS				
Privatisation	S.Claughton	NRA Proposal	EC/DWMJ	
HMIP	D.A.Mills	Deputy CI Wat & Wastes	JWL	
Rural Affairs	D.Brown	Countryside research	EC/CHM	
Planning	A.Ward	Minerals Planning	EC	
MINISTRY OF AGRICULTURE FISHERIES AND FOOD				
	D.Hardwick	All agricultural issues	EC/CHM,JWL	DWC Rodda
REGIONAL WATER AUTHORITIES, WATER COMPANIES ETC				
WA Association	R.White	Overview	EC/DWMJ	DWC Rodda,AC Skinner present at meeting
Yorkshire	Mr Franklin	(all major	EC	Dr J Aldrick, Dr Chadha
Anglian	A.Tetlow	issues: current	EC(also JWL at later meeting)	C Hayes, BT Croll
North West	R.Brassington	state of quality;	EC	
Severn Trent	A.Skinner	use;extent of	EC(also JWL at later meeting)	R Harris
Southern	H.Headworth	contamination;	EC	
South West	P.Chave	monitoring practice;	EC(also JWL at later meeting)	C Tubb
Thames	M.Morgan-Jones	future problems;	EC	R Flavin
Wessex	R Robinson	policy issues;	EC	DN Heath
Northumbrian	G.M.Kershaw	areas where research	EC	
Welsh	D.G.Walker	would contribute)	EC	W Logan Jack,P Bendry,S Eyre,B Buckley
Mid Kent	P.Bolas	Views & activities	EC	
Colne Valley	P.Medhurst	of the WC's	EC	
WELSH OFFICE				
WEP/Pollution Inspectorate	J.Atkins	Major issues as related to Wales	EC	J Saunders
NATURAL ENVIRONMENT RESEARCH COUNCIL				
NERC	Prof. Briden	Overview on research	EC	JD Mather,M Schulz,DWC Rodda
BGS	J.D.Mather	Specific research topics	EC/JWL	WM Edmonds,J Parker,P Whitehead,R Bradford (IH)
WATER RESEARCH CENTRE				
Medmenham	L.Clarke	Specific research topics	EC/JWL	
PROPERTY SERVICES AGENCY				
	R.Peet	Water Supply	EC	

1.4   Contents of the Report

The structure of this report is largely in accordance with the categorisation of issues for review as given in the terms of reference. A draft contents list for the Final Report was included in the Inception Report and no significant changes have been made to the proposed format.

The requirement to consider quantitative aspects of groundwater use only in so far as they have a bearing on quality issues was noted. However, it was considered early on in the project that some appreciation of the national and strategic importance of groundwater was required, together with the extent to which reliance is placed on certain aquifers, each with its particular characteristics. These findings are described in Section 2.

The quality of groundwater is described in Section 3, together with a review of the information on which the overall conclusions are based. This section includes a review of current groundwater quality monitoring practice and a description of natural hydrochemical conditions in the major aquifers.

Legislation aspects are discussed in Section 4, which includes the requirements of Acts of Parliament and the EC, together with less formal guidelines and codes of practice which affect groundwater quality. The requirements of current water quality standards are reviewed, together with the implications of privatisation.

The subject of research is discussed in Section 5, and includes a review of sources of funding and costs, current research and future requirements.

Section 6 presents views on future changes which are likely to affect groundwater quality based on presently available information. The review takes into account observed trends and possible changes in policy and legislation.

Section 7 presents the conclusions drawn from information given in the report.

Section 8 contains priority recommendations relating to management practices for the protection of groundwater and issues requiring further more detailed examination, including research.

2. UTILISATION OF GROUNDWATER IN ENGLAND AND WALES

2.1   Introduction

The purpose of this chapter is to demonstrate the national and strategic importance of groundwater as part of the total water supply to England and Wales. National trends in groundwater abstraction are examined in terms of total water abstraction and use.

The reliance on groundwater by each of the ten regional water authorities is assessed and a breakdown of use is given. A regional and national assessment of the relative importance of each of the main aquifers in England and Wales is presented together with review of current groundwater resource management practices.

In preparation of this section the Consultants relied heavily upon data held by the Water Statistics and Policy section of the Water Authorities Association. Individual publications consulted are included in the list of references.

2.2   Water Supply and Demand

2.2.1   Total Water Supply and Demand
The total water supplied in England and Wales over the period 1975 to 1985 is shown in Table 1, categorised according to use. Table 1 is based on returns by licensed abstractors to each water authority.

Total water demand is shown in graphical form in Figure 2.1, which excludes abstraction by the CEGB, and for fish farming and watercress growing purposes. The CEGB abstract water for cooling, approximately 98% of which is returned to river systems close to the points of abstraction. Data on use for fish farms and watercress growing is incomplete because most abstractions for this purpose are unlicensed.

TABLE 1

WATER ABSTRACTION FOR ENGLAND AND WALES 1975 – 1985

Units in Ml/d

	Water Supply		Industry		Spray Irrigation		Agriculture		CEGB	Fish farming and watercress
	surf	grnd	surf	grnd	surf	grnd	surf	grnd		
1975	10294	5066	5188	1372	79	32	22	72	13714	-
1976	10142	4867	5328	1327	114	47	24	72	13211	-
1977	9985	4762	5603	1355	85	31	25	95	13406	-
1978	10795	5033	5330	1296	60	21	41	109	12539	-
1979	11082	5185	5462	1300	74	32	33	106	12710	-
1980	11052	5134	3797	1237	54	38	31	102	10474	-
1981	10796	5309	3695	1278	72	45	9	102	9284	-
1982	11062	5269	3632	1097	91	48	8	109	8931	970
1983	10925	5299	3095	998	108	63	6	112	8254	971
1984	10987	5415	2952	940	125	74	9	113	7052	1066
1985	11347	5294	2958	981	64	43	12	118	4369	1105

* excluding water power (hydro-electric), saline and tidal abstractions

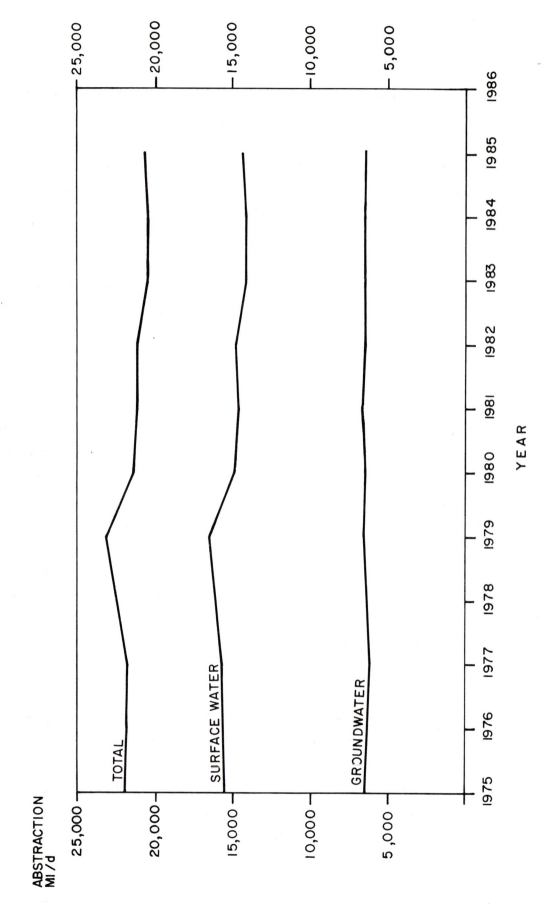

Figure 2·1

TOTAL WATER SUPPLY 1975-1985 IN
ENGLAND & WALES

Note: Excludes CEGB, Fish Farming & Water Cress Growing Abstractions

Between 1975 and 1985 total demand declined from 22125 Ml/d to 20820 Ml/d. The decline commencing in the late 1970s may be attributed in part to the success of measures to reduce wastage following the drought in 1976. However as discussed later in 2.2.3 it may also be due to changes in industrial use. Data from recent years indicate a slight upturn in demand.

The data include only licensed abstractions, which cover the vast majority of abstractions; there are numerous minor abstractions, chiefly for general agricultural purposes, that are exempt from licensing and for which no records are kept.

2.2.2   Public Water Supply

Water for public supply in England and Wales is abstracted under licence by water undertakings which comprise the 10 regional water authorities and 28 water companies. Most of the water is supplied after treatment for domestic use with a smaller proportion supplied for industrial and agricultural purposes.

The historical growth of public supply for England and Wales from 1901 is shown in Figure 2.2. The main features and trends are:

o   a steady annual rise in demand from 1901 to 1941 of 70 Ml/d;

o   a distinct increase between 1941 and 1971 when demand increased on average by 270 Ml/d each year;

o   a reduced average annual increase of 160 Ml/d since 1971; the reduction may be explained by more efficient use of water during the 1970s;

Figure 2·2

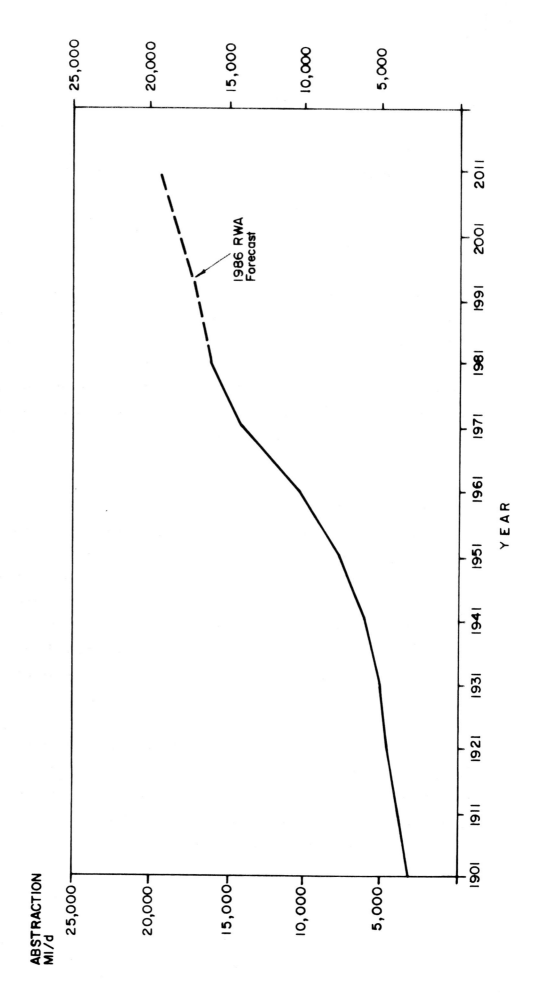

TOTAL PUBLIC WATER SUPPLY FROM 1901 IN
ENGLAND & WALES

o   between 1980 and 1985 (the most recent year for which
         figures are available) abstraction for public water supply
         was in the range 16,000 to 17,000 Ml/d, approximately 75%
         of the total abstraction, under licence, excluding
         abstraction by the CEGB.

2.2.3   Private Industrial and Agricultural Water Supplies

Abstraction for private industrial and agricultural purposes is
shown in Table 1 and is distinct from public water supplies in
that the user directly abstracts water under licence. As stated in
Section 2.2.2 most industrial and agricultural supplies come from
the public water undertakings but because there are no readily
available figures for these amounts the discussion on trends in
industrial and agricultural usage refers to private abstractions
only.

Since the late 1970's there has been a 40% decrease in privately
licensed industrial usage of water. The decrease is due to three
main factors, (i) to a decline in traditional water-using
manufacturing industries, (ii) to a more efficient use of water
and (iii) to a move towards metered supplies from the public water
undertakings. In recent years the private industrial use of water
appears to be stabilising; in 1985 abstraction amounted to
3939 Ml/d which represented 18% of the total licensed water
demand.

Trends in agricultural use are not so clear. Abstraction for spray
irrigation varies signficantly from year to year depending on the
weather during the growing season, though the trend appears to be
upward. Abstraction for other agricultural purposes, chiefly
livestock watering, has been relatively constant. In 1985

total licensed abstraction for the agricultural industry amounted to 237 Ml/d which is 1% of the total licensed water demand.

It is estimated that the water demand from the agricultural industry in England and Wales is 700 Ml/d which is approximately three times the quantity privately abstracted from groundwater and surface water sources under licence. The difference is made up from:

(i)   Supplies from public water undertakings (generally in lowland areas);

(ii)   Livestock, particularly in the uplands, obtaining water from natural surface sources.

Other private uses include fish farming and water cress growing for which the abstraction under licence in 1985 was 1105 Ml/d or 5% of the total. There are insufficient data to indicate any trends.

2.2.4   Summary

Total water demand in England and Wales appears to be increasing again following a decrease in the late 1970s - early 1980s which was due to more efficient use in water and a reduction in industrial use. The increase is occurring mostly in the public water supply sector which in recent years accounted for some 75% of the total licensed water supply. Privately licensed industrial abstraction, after a recent decline, now appears to be stabilising. Agricultural use seems to be rising due mainly to increasing spray irrigation.

2.3   Groundwater Abstraction and Use

2.3.1   National Trends

Data on groundwater abstraction are available from 1948, following the Water Act of 1945 which required returns to be submitted from public water undertakings and private users. Section 114 of the 1963 Water Resources Act empowered the 29 newly-formed river authorities to obtain abstraction information from surface water and groundwater users. However records between 1964 and 1972 are either incomplete or unreliable.

Groundwater abstraction accounts for approximately 30% of the total water demand in England and Wales, excluding abstractions by the CEGB. A breakdown of the total water supply into groundwater and surface water for the period 1975 to 1985 is shown in Table 1 and as a graph in Figure 2.1; the contribution from groundwater has remained relatively constant: 29% in 1975 rising slightly to 31% in 1985.

Total groundwater abstraction, with a breakdown into public supply and private industrial supply categories, is shown graphically for the period 1948 to 1985 in Figure 2.3. From 1948 to the early 1970s groundwater abstraction increased steadily from 4250 Ml/d to 6750 Ml/d due solely to the increase in the demand for public supply; private industrial demand remained relatively constant. Since the early 1970's total groundwater abstraction has been reasonably steady at approximately 6500 Ml/d balanced by the continuing increase in public supply and the decrease in privately licensed industrial supply. For reasons given in 2.2.3, the decline in private industrial use has meant that as a proportion of total groundwater abstraction, groundwater abstracted by public supply undertakings has risen from 66% in 1948 to 82% in 1985.

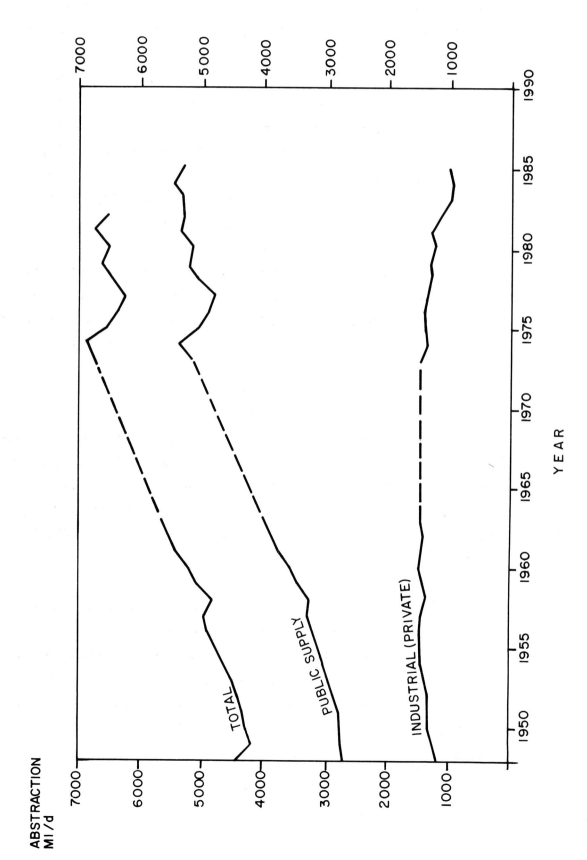

Figure 2·3

Currently, groundwater privately abstracted under licence for agricultural purposes amounts to approximately 3% of the total groundwater abstraction. The data given in Table 1 show that:

o there is a significant annual variation in the quantity of water required for irrigation depending on the weather during the growing season. For example the demand in 1985 was almost half that of the preceding year. Generally though, the demand for groundwater for irrigation is increasing and this is attributed specifically to the east of England where surface water availability is lower;

o in contrast to irrigation, the demand on groundwater from other agricultural activities has been steady over the period 1978 to 1984, although the use of groundwater is far greater than surface water usage in this category.

2.3.2   Regional Water Authorities

Groundwater abstraction in 1985 in each of the ten regional water authorities in England and Wales is shown in Table 2 to illustrate the present situation. As illustrated in Figure 2.4 most groundwater abstraction takes place in the south and east of England corresponding to the distribution of the major Chalk aquifer.

In volumetric terms, groundwater abstraction was greatest in the Thames Water area where it reached 1800 Ml/d in 1985. In terms of percentage of total water supply, Southern Water Authority relied most heavily on groundwater which amounted to 73.6% of the total in 1985. In contrast groundwater accounts for 10% or less of the total water supply in the Welsh and Northumbrian Water Authorities.

TABLE 2

GROUNDWATER ABSTRACTION IN THE REGIONAL WATER AUTHORITIES: 1985

		ANGLIAN	NORTHUMBRIAN	SEVERN TRENT	NORTH WEST	SOUTHERN	SOUTH WEST	THAMES	WELSH	WESSEX	YORKSHIRE
TOTAL ABSTRACTION	Ml/d	1946	1077	2710	3656	1353	655	4160	2468	894	1898
GROUNDWATER ABSTRACTION	Ml/d	986	110	1063	471	996	114	1802	138	408	348
GROUNDWATER ABSTRACTION AS % OF TOTAL		50.7	10.2	39.2	12.9	73.6	17.4	43.4	5.6	45.6	18.3
% BREAKDOWN OF GROUNDWATER ABSTRACTION BY USE		77									
PUBLIC SUPPLY		19	82	84	60	88	58	90	60	88	76
PRIVATE SUPPLY:											
INDUSTRIAL		2	18	15	39	11	14	9	30	6	21
SPRAY IRRIGATION		2	-	0.5	n	n	1	n	1	n	1
AGRICULTURE (OTHER)		2	-	0.5	1	1	27	1	9	6	2

Notes: (i) n = negligible
(ii) excludes abstraction by CEGB

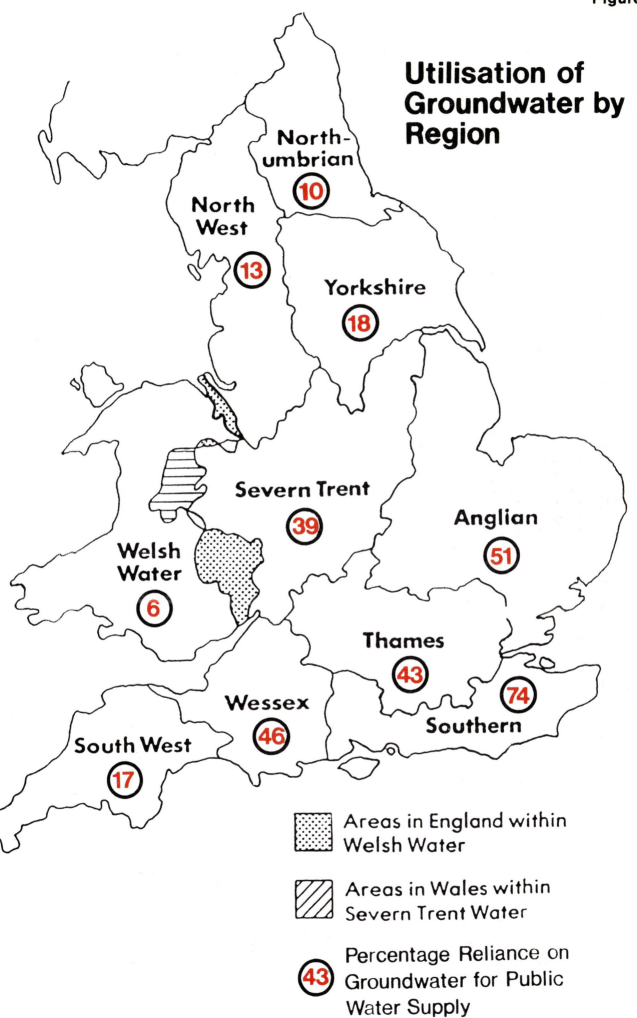

Figure 2.4

Utilisation of Groundwater by Region

North-umbrian (10)

North West (13)

Yorkshire (18)

Severn Trent (39)

Anglian (51)

Welsh Water (6)

Thames (43)

Southern (74)

Wessex (46)

South West (17)

▨ Areas in England within Welsh Water

▨ Areas in Wales within Severn Trent Water

(43) Percentage Reliance on Groundwater for Public Water Supply

Generally some 70% to 90% of the total groundwater abstraction was used for public water supply purposes (see Table 2). However there were three notable exceptions: in the North West and Welsh Water Authority areas private industrial usage of groundwater assumed greater significance, whilst in the South West Water Authority, 30% of groundwater abstracted was used privately for agriculture.

2.3.3   Groundwater Contribution to Surface Water

In many major river systems in England and Wales a large proportion of the flow during the summer months is baseflow, originating as groundwater from the major and secondary aquifers. Thus for part of the year surface water abstractions may be mainly groundwater fed.

In addition several schemes have been built to enable river regulation by groundwater during droughts. The status of the main schemes is summarised in Table 3. In general, although these facilities are operational they have rarely been used because there have been no significant droughts since 1976 to cause sufficiently low river flows.

2.4   Groundwater Abstraction by Aquifer

Actual abstraction in 1977 from each of the main aquifers in each regional water authority area in England and Wales is given in Table 4. The distribution of the main aquifers is shown on Figure 2.5 and percentage aquifer utilisation is summarised on Figure 2.6. Of paramount importance is the Chalk aquifer which provided 54.5% of the total groundwater supplies. Second in importance are the Permo-Triassic sandstones which provided 26.0% of the total. No single remaining aquifer provided more than 5% of the total groundwater abstraction.

TABLE 3: RIVER REGULATION USING GROUNDWATER - MAJOR SCHEMES IN ENGLAND AND WALES

WATER AUTHORITY	SCHEME	AQUIFER	RIVER	COMMENTS
Thames	Thames Groundwater Scheme	Chalk (major) + Jurassic Oolites (minor)	Thames	Scheme operational but final operational rules have yet to be decided - awaits minimum acceptable flow at Teddington to be set. Scheme so far used in 1976 only.
Severn - Trent	Shropshire Groundwater Scheme	Sherwood Sandstone	Severn	Scheme consists of 8 stages. First stage complete, second stage under construction, completion of all stages by 2011. First stage used in 1984 only (capacity 34Ml/d) and not since. Second stage will have capacity of 32Ml/d
Anglian	Great Ouse Scheme	Chalk	Great Ouse & Tributaries	Scheme is part operational but up to now has had only minor seasonal use; has never been heavily used as river flows have generally not been low enough.
Southern	Candover Scheme	Chalk	Itchen	Scheme commissioned in 1976 (pilot tested that year) - since then has not been put into use as river flows have not been sufficiently low. Capacity of scheme is about 55Ml/d.

TABLE 4: GROUNDWATER ABSTRACTION BY AQUIFER IN ENGLAND AND WALES IN 1977

WATER AUTHORITY

GEOLOGICAL ERA	AQUIFER	ANGLIAN Ml/d	%	NORTH WEST Ml/d	%	NORTHUMBRIAN Ml/d	%	SEVERN TRENT Ml/d	%	SOUTHERN Ml/d	%	SOUTH WEST Ml/d	%	THAMES Ml/d	%	WESSEX Ml/d	%	YORKSHIRE Ml/d	%	WELSH Ml/d	%	TOTAL Ml/d	% of TOTAL
DEVONIAN	OLD RED SANDSTONE																			13	5	13	0.2
CARBONIFE-ROUS	CARBONIFEROUS LST			24	4			13	1							101	16	6	2	125	51	269	4.2
	MILLSTONE GRIT			23	4													13	5	2	1	38	0.6
	COAL MEASURES*			25	4	1	1	61	6									41	17	43	18	171	2.7
PERMIAN	MAGNESIAN LST					85	89	10	1									18	7			113	1.8
PERMO-TRIAS	PERMO TRIASSIC SANDSTONES	512	88			9	10	939	92			45	80			3	1	89	36	61	25	1658	26.0
JURASSIC	MIDDLE: LINCS LST	117	13																			117	1.8
	: OOLITIC LSTs	6	1											52	3	112	18	6	2			176	2.8
	UPPER: CORALLIAN																	35	14			35	0.5
CRETACEOUS	LOWER:SPILSBY SST	23	3																			23	0.4
	:HASTINGS BEDS									51	6											51	0.8
	:LOWER GREENSAND	45	5							70	8			122	7							237	3.7
	UPPER:CHALK (&UGS)	679	78							746	86	11	20	1588	90	400	65	42	17			3466	54.5
	TOTAL	870		584		95		1024		867		56		1762		616		250		244			

GRAND TOTAL 6367

* NOTE: Figures for Coal Measures from 1972/73

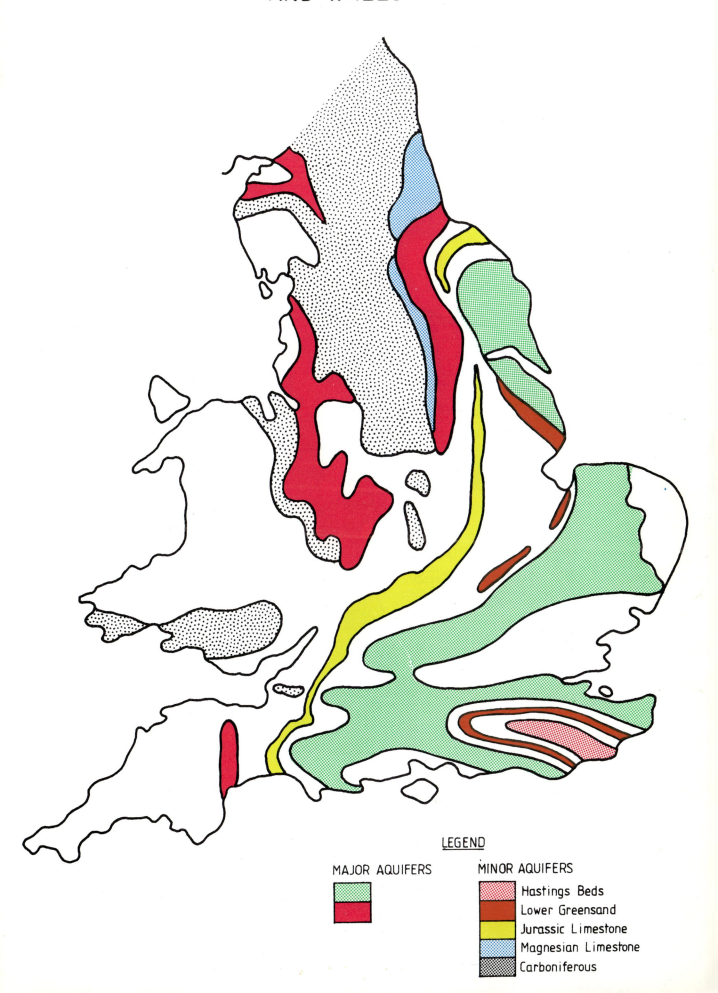

Figure 2.5

AQUIFERS OF ENGLAND AND WALES

LEGEND

MAJOR AQUIFERS

MINOR AQUIFERS

Hastings Beds
Lower Greensand
Jurassic Limestone
Magnesian Limestone
Carboniferous

Figure 2.6

GROUNDWATER ABSTRACTION IN 1977 IN ENGLAND & WALES BY AQUIFER

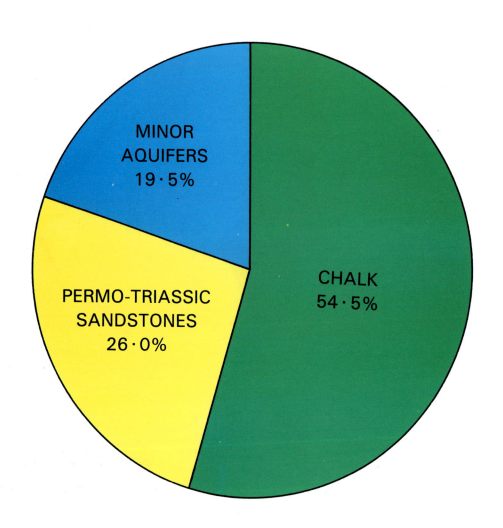

Halcrow

The main points regarding aquifer utilisation within the regional water authorities are:

o   In the Thames, Southern, Anglian and Wessex Water Authority areas the Chalk is the aquifer which is primarily utilised for groundwater supplies; utilisation is greatest in Thames Water Authority with 90% of groundwater coming from the Chalk.

o   In the Severn Trent, North West, and South West Water Authorities Permo-Triassic sandstones are the main aquifers utilised and provide over 80% of the total groundwater within the three authority areas.

o   Of the remaining three water authorities, several aquifers are utilised in Yorkshire (the Permo-Triassic sandstones having the highest percentage), whilst the main aquifers in the Welsh and Northumbrian Water Authority areas are the Carboniferous Limestone and Magnesian Limestone respectively.

Although the above figures are based on 1977 data, information provided by the water authorities during this study indicates that there have been no significant changes in the proportional use of aquifers since 1977.

Abstraction from the Coal Measures warrants a special mention. Most groundwater abstracted from the Coal Measures originates from mine drainage activities and currently such abstractions are exempt from licensing. Prior to 1963 the NCB submitted returns on quantities of groundwater abstracted for mines drainage under Section 6 of the Water Act, 1945. However the only data since then were obtained from a survey carried out by the NCB in 1972/73 which showed that 870 Ml/d were discharged from coal mines, 85% of which came from three coalfields: Northumberland and Durham, South Wales and Yorkshire.

In 1973 minewater discharge amounted to approximately 13% of the total groundwater abstraction in England and Wales but since only about 30 Ml/d are used for supply purposes and the remainder is discharged to rivers (97%), minewater drainage from the Coal Measures is excluded from the statistical analysis of abstraction from the main aquifers given above. However Table 3 does include 1972/73 groundwater abstraction data for borehole water supplies from the Coal Measures.

Data for abstraction from widespread but minor sand and gravel aquifers found in deposits of Tertiary and Recent age are not available from every water authority and therefore have not been included in Table 4. It has been estimated that these aquifers nationally contribute some 3.5% of the total groundwater abstraction. This percentage rises to 7% for the Severn Trent and North West Water Authorities and 6% for the Thames Water Authority.

2.5   The Importance of Groundwater

The data reviewed in this section indicate that approximately 30% of drinking water in England and Wales is abstracted as groundwater, though reliance on groundwater by individual water authorities varies from less than 6% to nearly 74%. Nevertheless even within the water authorities which are small groundwater users there exist centres of population which are heavily or totally reliant on groundwater as a source of supply. Providing alternative sources to such locations would be very expensive.

16

In addition, as described earlier, in some rivers from which water is abstracted, the flow during the summer months is mainly groundwater, originating either as natural baseflow or occasionally pumped as part of an augmentation scheme. As such a proportion of surface water abstraction is in effect groundwater for a significant period of time in the summer months, further emphasising the importance of groundwater as a source of supply.

3.   THE QUALITY OF GROUNDWATER

3.1   Introduction

This section contains an appraisal of the current state of
groundwater quality in England and Wales and the factors which
affect it.

Present practices for monitoring groundwater quality are described
together with a review of the natural hydrochemical conditions in
the major aquifers. A review of the principal sources and types of
contamination is given based on research carried out during the
course of the project, and is followed by an overview of the state
of groundwater quality.

3.2   Groundwater Quality Monitoring

3.2.1   Purpose and Responsibility

Monitoring of the quality of groundwater used for public
consumption is mainly carried out by the regional water
authorities and water companies. Monitoring may be carried out by
one of several divisions or groups into which the individual
authorities are divided, including Operations, Technical and
Scientific Services, Planning and Engineering.

The following purposes for which monitoring is carried out have
been identified:

(a)   To ensure compliance with EC Water Quality Directives
       currently in force;

(b)   To monitor the quality characteristics of raw groundwaters in order to ensure the appropriateness of treatment processes required to provide a wholesome supply;

(c)   As part of a groundwater resources management programme to ensure long-term conservation of the resource;

(d)   In hydrogeochemical studies forming part of groundwater investigations for additional sources;

(e)   As part of groundwater pollution investigations, in the form of ad-hoc sampling and analysis for specific substances which may originate from site-specific or diffuse sources;

(f)   As part of research projects, which most frequently relate to groundwater pollution studies.

Legislation affecting groundwater quality is reviewed in Section 4, but is introduced here in order to describe monitoring activities in the context of responsibility and purpose.

The EC Drinking Water Quality Directive 80/778/EEC (Department of the Environment, 1982) specifies the quality of water used for public consumption, and includes the requirements for sampling and analysis. However, this Directive concerns water as it reaches the consumer and there is no requirement to sample and analyse the raw groundwater. Purpose (a) in the list above therefore relates to treated water while purpose (b) is essentially an operational requirement to ensure that the standards specified in the Directive are met.

There is at present no British legislation specifying the quality standards of raw groundwater or a requirement to monitor it.

However it might be argued that the responsibility of the water authorities in this repect is implied in Clause 10 of the Water Act 1973 where they are required to:

"take all such actions as (they) may from time to time consider necessary or expedient ... for the purposes of conserving ... water resources in their area ..."

Purpose (c) is most relevant to this aspect of the 1973 Water Act as, to some extent are purposes (d) and (e). Groundwater quality monitoring is most effective when it is carried out as part of a total groundwater resources management strategy, from which an appreciation of the long-term implications of water quality issues can be gained. Hydrogeochemical studies forming part of new groundwater resources investigations are particularly important in this respect as the opportunity arises for establishment of baseline values, against which changes identified during subsequent operational monitoring can be assessed.

Groundwater quality monitoring is also carried out as part of the investigations into pollution incidents or pollution control activities (purpose (e)). This is frequently in response to recognition of specific contamination problems, either observed or anticipated, of an urgent nature. EC Directive 80/68/EEC on the Protection of Groundwater Against Pollution Caused by Certain Dangerous Substances (Commission of the European Communities, 1979) is of relevance here because of the monitoring requirements required as a condition of authorisation.

However, taken out of its wider hydrogeochemical context such monitoring tends to be of value mainly in operational terms, that is, on the basis of the data obtained, supplies may be treated, blended, abandoned or maintained. As such the opportunity to

contribute to the understanding of long-term trends may be
overlooked in the light of the requirement to solve an immediate
problem.

Monitoring as part of research projects (purpose (f)) mainly
relates to pollution studies. The major research institutions (BGS
and WRC) have carried out a series of studies into various aspects
of groundwater pollution involving close collaboration with the
water authorities, who also undertake a limited amount of research
independently. Observation wells installed for research projects
may be monitored long term for specific substances or routine
chemical analysis, either by the research body or the water
authority.

3.2.2 Groundwater Quality Monitoring Practice

Groundwater quality monitoring practice varies significantly
between the regional water authorities. In most cases it is
treated groundwater which is monitored routinely and frequently
with sampling and analysis of raw groundwater being measured on an
infrequent or ad-hoc basis only. In addition, special monitoring
may be carried out around known or suspected sources of
contamination.

Routine monitoring is usually based on the requirements of the EC
Drinking Water Quality Directive, and incorporates biological,
inorganic and organic quality determinations. A manual describing
the monitoring practice is produced by some water authorities and
the results may be summarised in an annual water quality report.
Even in some water authorities which are heavy groundwater users
the monitoring, analysis and reporting is disproportionately
orientated towards surface water quality considerations.

21

In other authorities, each a major groundwater user, a network has been established to systematically monitor raw groundwater quality. In each case, the monitoring practice incorporates techniques based on recognition of the particular needs of groundwater sampling, which are discussed in the next section. In addition to the routine monitoring some pollution monitoring may be carried out at particular sites or more widely for specific substances.

Monitoring is also being carried out by research institutions either independently or in conjunction with the water authorities. This work includes site specific and regional studies, particularly for nitrate. The research institutions may also be called in to assist on other quality aspects of groundwater resources where insufficient specialist expertise exists within the authority.

Legislation on water quality increasingly necessitates detection of substances at very low concentrations. In addition, sampling to avoid cross-contamination and analytical procedures may be time-consuming, difficult and expensive. There are examples of this type of monitoring being carried out under contract by commercial organisations because of manpower constraints or lack of analytical facilities within the authority.

Two other organisations involved in groundwater quality monitoring for water supply purposes are Property Services Agency (PSA) and the Local Authorities. PSA are involved in monitoring of water quality on Crown Properties and the Local Authorities, through their Environmental Health Officer, monitor private wells. PSA monitoring is based on the requirements of the EC Drinking Water Quality Directive and mainly involves treated water, though some routine scans of the quality of raw groundwater are being carried out on Crown Properties by the Laboratory of the Government Chemist (LGC).

The situation with other private wells is not known, although in at least one water authority a network of private wells is included in the routine groundwater quality monitoring programme. These supplies are also governed by the EC Drinking Water Quality Directive and as such are concerned with water as supplied. However, it is likely that water as used by the consumer is frequently untreated and so data obtained from the monitoring of private wells could be a useful source of hydrochemical information.

3.2.3 Sampling Methods

An important aspect of water quality monitoring is the requirement for samples which when analysed provide data representing in-situ conditions, and hence provide a means by which the distribution and behaviour of contaminants can be properly understood. This is particularly difficult in the case of groundwater because of the capacity for physical and chemical change in waters as they move from the aquifer to the borehole and on into the sampling device.

The requirement for well-head determinations, fixing by acidification, multiple sampling etc is well established for major and minor ion suites but is a time consuming and expensive operation. Many extraneous factors need to be considered such as representative nature of open hole waters, mixing of waters from different horizons etc. However, with care and experience meaningful data is obtained and the understanding of natural hydrogeochemical distributions and processes in aquifers has increased significantly in recent years.

The problem of sampling for pollutants often at very low concentrations presents difficulties which at present have not been fully resolved. Certainly, the expense in terms of time and equipment for regional sampling where there is a requirement for unambiguous interpretation would under present circumstances be regarded as prohibitive in many cases. There is however no doubt that data of this accuracy are required if the distribution and behaviour of many substances currently regarded as troublesome is to be properly understood.

The subject of sampling is currently of concern to many individuals and institutions involved with groundwater quality. A review of some of the points at issue are:

o   the problem of mixing when sampling from open boreholes and the consequent difficulty of understanding 3-dimensional distribution of contaminants;

o   the requirement for flushing borehole sampling points, observed changes in concentration of substances with time at the same sampling installation;

o   the necessity to consider drilling methods and materials from which sampling equipment is made in relation to the substance sampled (particularly solvents);

o   the requirement for a statistical population of sample results for better interpretation;

o   the requirements for cheaper alternatives to currently available equipment;

o   consistency of conditions under which samples are taken
        (same borehole pumping at same rate for same duration);

o   the requirement for laboratories to cope with a small sample
        size when samples are taken from low yielding
        installations.

The need for research into sampling methods is discussed in
Section 5.

3.2.4   Archiving of Data

Systems operated by the authorities for storage and retrieval of
groundwater quality data also vary. Data may be stored on
sophisticated computer systems or archived manually as printed
records.

The computer systems usually have available the useful facilities
associated with data base storage and retrieval, including methods
of highlighting when parameter measurements exceed a certain
value. Data can also be manipulated to provide graphical plots to
illustrate trends in concentration with time for specific
substances. However the systems operated by individual water
authorities are not necessarily compatible with those of others
and there is currently no national system for archiving
groundwater quality data obtained.

No comprehensive national archive of water quality data is
operated by the research institutions, other than that associated
with specific projects such as, for example, nitrate studies and
acid rain. The national network of observation wells now operated
by the Institute of Hydrology (formerly by the Water Data Unit)
does not include water quality determinations.

3.2.5   Cost of Monitoring

No clear picture has emerged on the cost of groundwater quality
monitoring. The different approaches to monitoring operated by the
water authorities make it very difficult to derive a cost
applicable to groundwater sampling and analysis alone.

Factors involved in sampling and analysis of groundwater are
essentially as follows:

o   sampling facility (well, piezometer etc)

o   sample collection (manpower and equipment)

o   analysis (manpower equipment and consumables)

o   storage and retrieval of data

The sampling facility may be purposely constructed for monitoring
purposes or be a water supply borehole. Similarly, manpower and
equipment costs associated with sample collection, analysis of
groundwater samples and archiving of data, may form part of other
quality monitoring activities.

The costs of sampling and analysis for many substances for which
straightforward analytical techniques are available do not appear
to be a problem as far as the water authorities are concerned.
However concern has been expressed regarding the increasing
requirement for monitoring a series of complex organic compounds
such as pesticides and solvents at very low concentrations. Such
analyses frequently necessitate use of GCMS facilities which not
all authorities have. One water company which is very heavily
reliant on groundwater sources reported that £90,000 was due to be

spent in new laboratory equipment, much of which was to meet the requirements of the EC Drinking Water Quality Directive (op.cit).

3.3 Natural Hydrochemical Conditions

3.3.1 Introduction

The natural hydrochemistry of groundwater is determined primarily by the interaction of (i) recent, meteoric water, (ii) in situ, older groundwater and (iii) the mineralogy of the aquifer. As a result groundwater chemistry normally comprises a combination of the following major ions:

Cations: Calcium (Ca), Magnesium (Mg), Sodium (Na), Potassium (K), and to a lesser extent Iron (Fe)

Anions: Carbonate (CO_3), Bicarbonate (HCO_3), Sulphate (SO_4), Chloride (Cl)

In most aquifers in England and Wales, particularly in the recharge zones at outcrop, groundwaters are characterised by hardness. Total Hardness is made up of carbonate (temporary) hardness caused by calcium and magnesium bicarbonate, and non-carbonate (permanent) hardness caused by other salts of calcium and magnesium. Due to the prevalance of calcareous aquifers in England and Wales carbonate hardness normally dominates.

Natural changes occur within an aquifer during groundwater flow which alter its hydrochemical character. Softening of groundwater, where calcium and magnesium are exchanged for sodium ions, reducing conditions, where dissolved oxygen decreases and nitrate and sulphate are reduced, and mixing with old or connate groundwaters, with a resulting increase in salinity, are common features in many aquifers.

The natural groundwater chemistry of the major aquifers in England and Wales is described in this section and the significant variations are highlighted. Figure 2.5 is a map showing aquifer locations.

3.3.2   The Chalk

The Chalk aquifer occurs extensively in England, south east of a line between Yorkshire and Wessex.

Chalk groundwater is typically hard in the aquifer at outcrop. Total hardness generally varies between 200 and 300 mg/l, 80% to 90% of which is carbonate hardness. Calcium bicarbonate is dominant due to the almost wholly calcareous nature of the rock. Groundwater both above and below the water table is often saturated with respect to calcite. Natural chloride concentrations are low, typically 20 mg/l or less.

Groundwater flow in the Chalk aquifer follows three main courses:

.   discharge to rivers, notably the chalk streams of Hampshire
      and the winterbournes of the Chiltern Hills

.   direct discharge to the sea where the aquifer meets the
      coast, for example in Sussex

.   flow from the recharge zones into the confined aquifer
      beneath the Tertiary clay cover, either naturally or under
      the influence of pumping, for example, in the London Basin.

Only in the third case do significant natural changes in Chalk groundwater chemistry occur. Groundwater softening, reduction of oxygenated compounds and mixing with connate water have all been identified as processes occuring in the confined Chalk aquifer of the London Basin and East Anglia. For example 13 km into the confined aquifer in North Kent total hardness decreases to 25 mg/l and sodium bicarbonate replaces calcium bicarbonate as the major compound as a result of the softening (or ion exchange) process. The reducing conditions in the confined aquifer is caused by lack of dissolved oxygen and reduction of nitrate occurs. Also, as a result of mixing with old, possibly connate, groundwater, chloride concentrations increase further into the confined zone. For instance in Berkshire a peak in chloride concentration of 100 to 200 mg/l occurs in the centre of the London Basin syncline. In East Anglia chloride concentrations are higher, up to 1000 mg/l in the confined aquifer beneath the Essex coastline.

3.3.3   The Permo-Triassic Sandstones

The Sherwood Sandstone aquifer, newly named to include the Bunter Sandstone, is the principle aquifer of Permo-Triassic age and outcrops extensively throughout the Midlands, the North West and South West of England.

There are regional variations in groundwater chemistry of the Sherwood Sandstone beneath outcrop due to lithological changes within the aquifer but, as with the Chalk, the major changes occur at depth or where flow is from unconfined to confined conditions. The principal hydrochemical characteristics of the Sherwood Sandstone aquifer are presented below for individual major source areas.

a)   East Midlands and Yorkshire

In groundwater beneath the outcrop the dominant compound is
calcium bicarbonate and hardness is medium to high. The highest
quality waters in terms of potability are located east of outcrop
along the western margin of the confined area where softening
takes place and chloride concentrations are exceptionally low (ie
less than 10 mg/l). The low chloride values have been attributed
to older groundwaters derived from less saline continental-type
rainfall unaffected by more recent and more saline maritime type
rainfall and also unaffected by anthropogenic activity. Further
into the confined zone, however, the groundwater quality
deteriorates due to a significant increase in sulphate as a result
of dissolution of gypsum and anhydrite beds; this change occurs
some 15 to 25 km east of the outcrop, and the groundwater is not
potable.

b)   Shropshire

Saline waters occur at depth (greater than 200 metres), beneath
outcrop and have been interpreted as old, possibly connate water
beneath a recent meteoric water circulation zone and also by
leakage of mineralised groundwater originating from contiguous
Carboniferous strata. Sulphate concentrations can be locally high
due to dissolution of gypsiferous beds.

c)   Worcestershire & Warwickshire

In contrast to the East Midlands, good quality groundwater occurs
within the confined aquifer and natural softening has resulted in
a sodium sulphate type groundwater, notably in the vicinity of
Stratford upon Avon.

d)   Staffordshire and Derbyshire

Groundwater quality is generally good, being of a calcium
bicarbonate type. Chloride concentrations tend to be enhanced in
proximity to the Coal Measures. In Derbyshire, groundwaters are
characterised by enhanced fluoride levels.

e)   Cheshire and Lancashire

Within the Cheshire basin natural brines are found at depth within
the aquifer as a result of groundwater dissolution of halite at
the base of the overlying Mercia Mudstones. In Lancashire, high
mineralisation of groundwater occurs also as a result of
dissolution of halite beds in the confining Mercia Mudstone.

f)   South Devon

Compared to outcrops further north the Sherwood Sandstone in South
Devon (known locally as the Budleigh Salterton Pebble Beds and
Otter Sandstone) is relatively drift-free and decalcified due to
weathering. As a result groundwaters have tended to be more acidic
with pH in the range 5 to 6. Total dissolved solids are low, less
than 150 mg/l and alkalinity is generally less than 25 mg/l.
However there are problems with naturally ferruginous groundwaters
which in the past have caused corrosion of well linings; the
situation has now been remedied and the problem is under control.

Where the Sherwood Sandstone is covered by drift deposits,
particularly in the Midlands, any recharge reaching the
groundwater through the drift tends to enhance concentrations of
magnesium and sulphate in the groundwater.

3.3.4   Lower Cretaceous

a)   Lower Greensand

The Lower Greensand aquifer (comprising the Hythe and Folkestone Beds) outcrops in a narrow band between the Wash and Bedfordshire and also rings the Wealden anticline in Kent, Sussex and Surrey.

Beneath outcrop, the hardness of groundwater varies with the degree of cementation but in general there is undersaturation with respect to calcite. In Kent, for example, total hardness of the Folkestone Beds aquifer varies between 20 and 350 mg/l. The hardness of Hythe Beds groundwater increases in an easterly direction as the calcareous content of the rock increases; total hardness is generally greater than 100 mg/l. Hythe Beds groundwater in Sussex is somewhat less hard than in Kent.

Generally, chloride concentrations in groundwater beneath Lower Greensand outcrop are low, typically 20 mg/l.

In the Folkestone Beds aquifer in North Kent there is significant groundwater flow in the confined zone beneath the Gault Clay. Associated with groundwater flow are chemical changes such as softening which reduces total hardness to 50 mg/l at Gravesend. In addition reducing conditions result in a decrease in dissolved oxygen content and, due to mixing, chloride concentrations increase up to 50 mg/l.

Lower Greensand groundwaters frequently contain naturally high concentrations of iron due to the often ferruginous nature of the sand. Folkestone Beds groundwater is particularly prone to high iron content; iron concentrations as high as 10 mg/l have been recorded.

b)   Hastings Beds

The Ashdown Sandstone and Tunbridge Wells Sandstone, forming part of the Hastings Beds, are present as minor aquifers in the Weald of Kent and Sussex.

Groundwater in the Ashdown Sandstone exhibits a wide variation in hardness, from relatively soft to values in excess of 200 mg/l. Natural softening occurs in the confined aquifer beneath the cover of Wadhurst Clay. Ashdown sandstone groundwaters are characteristically highly ferruginous because ironstone horizons are common; iron concentrations up to 13 mg/l have been recorded. In association with iron, manganese can be high, up to 3 mg/l.

Groundwater in the Tunbridge Wells Sandstone is generally soft with total dissolved solids in the range 100 to 175 mg/l, but can be as high as 300 mg/l. Chlorides are usually less than 20 mg/l.

c)   Spilsby Sandstone

The Spilsby Sandstone outcrops locally in Lincolnshire.

At outcrop total hardness ranges between 150 and 300 mg/l and chloride concentrations are typically 20 mg/l. In conjunction with regional eastward flow of groundwater in the aquifer the groundwater undergoes softening and chloride concentrations increase to above 250 mg/l in the vicinity of Skegness on the Lincolnshire coast.

3.3.5   Jurassic

a)   <u>Lincolnshire Limestone</u>

As its name implies the Lincolnshire Limestone aquifer refers to
an outcrop and confined area of Jurassic Limestone within
Lincolnshire. The hydrochemical changes in the groundwater on its
eastward passage from outcrop into the confined zone have been
well documented by a series of investigations. The principal
characteristics are described below.

Calcite saturation occurs in the recharge water migrating through
the unsaturated zone. Little change occurs until 10km into the
confined zone; up to this point the groundwater is predominantly
of calcium bicarbonate type. A marked redox barrier is present at
10km: a reduction in the redox potential marks the onset of
reducing conditions where oxygen decreases and nitrate is removed
by denitrification. Sulphate is also reduced to hydrogen sulphide,
which is however only present in trace amounts. Further east there
is a mixing of relatively recent groundwater with older, possibly
connate, groundwater with a corresponding increase in sodium and
chloride concentrations to high levels.

Fluoride concentrations south and west of Boston are noticeably
high and range between 1 and several mg/l.

b)   <u>Great and Inferior Oolitic Limestones</u>

The Great and Inferior Oolitic Limestones form the Cotswold Hills
running from the East Midlands through to the West Country.
A hydrochemical survey of the two aquifers has shown lateral
changes from unconfined to confined conditions similar to those
described above for the Lincolnshire Limestone.

At outcrop the groundwater is dominated by calcium bicarbonate. Sodium bicarbonate increases due to softening in the confined aquifer beneath the Oxford Clay. A redox barrier is evident at 8km into the confined Inferior Oolitic Limestone aquifer and at 1 to 2km into the confined Great Oolitic Limestone aquifer.

3.3.6   Permian (Magnesian Limestone)

The Magnesian Limestone aquifer outcrops in a north-south direction between Durham and Nottinghamshire. Total hardness of the groundwater is variable from hard to very hard. For example the range for the Upper Magnesian Limestone is 300 to 700 mg/l (higher values occur nearer the east coast) and for the Lower Magnesian Limestone the range is 150 to 400 mg/l. High values for hardness have restricted development of groundwater supplies.

3.3.7   Carboniferous

Principal outcrop areas of Carboniferous strata are the North of England, the Midlands, South Wales and the Bristol area.

a)   Carboniferous Limestone

Generally, Carboniferous Limestone groundwaters are very hard and saturated with respect to calcite. Concentrations of trace elements such as fluoride and lead are often enhanced (compared to other aquifers) near to mineralised zones in the rock.

In South Wales the Carboniferous Limestone represents the main aquifer; supplies are chiefly obtained at spring discharge points. Total hardness lies in the range 250 to 300 mg/l, and chloride concentrations are generally less than 30 mg/l. Due to the karstic nature of the aquifer, heavy rainfall events can cause turbidity problems for water supply sources.

b)   Millstone Grit

Being a sandstone formation with low calcareous content groundwaters within the Millstone Grit are often very soft. In South Wales total dissolved solids are typically below 200 mg/l.

c)   Coal Measures

Groundwaters contained in the Coal Measures are generally hard but the hydrochemistry may be very variable. Iron concentrations are often high, particularly in mining areas and chlorides commonly reach high levels. However chloride levels in Coal Measures groundwater in South Wales are exceptionally low and consequently the water is sometimes potable, although at depth the quality deteriorates with low pH and high iron, sulphate and total dissolved solids (up to 10,000 mgl/). In the Coal Measures of North East England the groundwater quality deteriorates coastwards; total dissolved solids rise to several thousand mg/l.

3.3.8   Devonian (Old Red Sandstone)

The Old Red Sandstone aquifer is only developed for groundwater resources to any extent in Wales. The groundwater is hard, typically levelling 220mg/l, 90% of which is carbonate hardness. Total dissolved solids are generally less than 300 mg/l.

3.4   Sources and Types of Contamination

3.4.01   Introduction

This section contains a review of the principal forms of contamination currently affecting or constituting a threat to groundwater quality in England and Wales.

Sources of contamination are described according to activity, for example, agricultural activities, waste disposal, etc, and the kinds of pollution which may result.

3.4.02   Pollution from Agricultural Sources

Agricultural activities are a significant source of groundwater contamination over wide areas of England and Wales. Problems with specific substances have been identified, particularly nitrate, and have been the subject of a considerable amount of research in recent years. Agricultural pollution is therefore considered to be a particularly important subject and is reviewed in more detail in Appendix C.

A summary is presented below:

a.   Background

Since the war there has been a continuous increase in agricultural productivity and production and as a result the United Kingdom is now more than self sufficient in many temperate foodstuffs. This improvement in productivity owes much to technological advances, particularly over the last 10 to 15 years, due to research in such areas as improved crop varieties, more effective use of fertilisers, better control of pests and diseases, improved agricultural systems leading to a higher degree of specialisation and greater mechanisation. In addition, better education and training has led to improved management and hence a faster uptake of new technology.

Agricultural activity is the source of various forms of groundwater pollution. Unauthorised or accidental disposal of animal and silage effluent, releases from chemical and pesticide stores and disposal through soakaways have all occurred, sometimes with serious local consequences. However, these incidents are relatively minor compared with the impact of point source pollution from agriculture on surface water. To date, point source pollution of agricultural origin has not caused significant damage to groundwater resources. Diffuse pollution from agriculture however presents a far greater problem both in terms of its current and likely future impact on groundwater quality.

b.   Diffuse Sources

Nitrate

Agriculture is the major contributor to nitrate pollution in groundwater. The main factors affecting the level of nitrate are as follows:

o   Fertiliser Application Rates

        Typical leaching losses from winter cereals and spring sown
        cereals, root crops and oilseed rape are 40 and 50 per cent
        of the inorganic nitrogen applied respectively. Conversely,
        leaching losses from cut and grazed grass are thought to
        be 10 and 15 per cent respectively. Under these

circumstances leaching losses are a direct function of the quantity of inorganic nitrogen applied and are therefore highest in those areas where intensive arable cropping predominates.

o   Intensity of Stocking

Intensity of stocking also affects the level of leaching because the higher the number of livestock per unit area the greater is the quantity of organic nitrogen produced.

o   Timing of Nitrogen Fertiliser Application

During the late summer and early autumn, soil nitrate supply usually exceeds the uptake by arable crops. Consequently, the application of nitrogen fertiliser to most arable crops during this period will contribute directly to the reservoir of nitrogen in the soil organic matter.

o   Time of Planting

If no crops are planted in the autumn, all the soil nitrate released during this period will be leached. Conversely, early planting of autumn crops results in a greater uptake of nitrate by the crop.

o   Ploughing

Ploughing, particularly of permanent pasture, releases large quantities of nitrate which cannot be utilised by the succeeding crop.

o   Climate

The concentration of nitrate in groundwater is dependent
upon the mass of nitrate leached and the extent to which
this is diluted by rainfall. Therefore, for a given rate of
leaching the nitrate levels will be higher in areas of low
effective rainfall.

Nitrate in groundwater has been the subject of a considerable
amount of research in the last 10-15 years and significant
advances have been made in the understanding of its distribution
and behaviour.

Information received from water authorities broadly accords with
conclusions of the recent report by the Nitrate Coordination
Group (Department of the Environment, 1986). Nitrate levels in
many areas are rising though there are also reports of levels
stabilising or in some cases falling. Some borehole supplies are
above current EC Water Quality Directive limits and need to be
blended; there are few cases of wells being shut down because of
high nitrate levels. Where levels have fallen it is reported that
a relationship with reduced use of fertilizer applications can be
identified.

Pesticides

In general, farmers are likely to adhere closely to the
recommended application rates of pesticides because:

o   in most cases there is a fairly narrow application 'window'
        at which the pesticide is effective;

o   pesticides are expensive and farmers will tend to keep to
         the recommended application rates for this reason alone.

The potential for leaching of pesticides and the possible risks
incurred will largely depend on the following factors:

o   the proportion of the spray which settles on the soil;
o   the potential of the pesticides to volatilise;
o   the composition of the soil and sub-soil and their
         adsorption characteristics;
o   the degradation products and their ability to chemically
         recombine;
o   the biological activity of the degradation products.

Leaching of pesticides is therefore a very complex process and it
is often difficult to detect their presence in groundwater due to
their low levels of concentration, lack of knowledge regarding
degradation products and lack of adequate analytical techniques.

Most agriculturally used pesticides are applied to arable crops.
Consequently, the pesticide load is dominated by the agricultural
techniques adopted by farmers in the major arable areas in
England.

The present situation on occurrence of pesticides in groundwater
is not clear. Most water authorities report that pesticides have
been found but not all are monitoring on a routine basis. There
are also cases where the present limits are being exceeded. Given
the very low permitted limits under EC Directives the increasingly
widespread occurrence of pesticides is cause for serious concern.

c.   Point Sources

As indicated above, the impact of point source pollution from
agriculture is minor compared with that of pollution from diffuse
sources. The main sources of point source pollution are as
follows:

o   Pesticides - spraying of water courses, run-off, careless
    disposal of containers and washings together with
    spillages.

o   Sheep Dipping - disposal of sheep dips is a potential source
    of pollution though existing controls generally take account
    of these risks.

o   Animal and Silage Effluents - mostly contaminate surface
    water but inevitably some pollution can enter the
    groundwater system; isolated examples have been reported.

o   Farm Sewage Sludge - disposal of sludge may form a pollution
    risk but to date there have been no reported adverse effects
    on groundwater.

3.4.03   Landfill Sites

Landfill sites are a common form of waste disposal and are
therefore widespread actual or potential source of groundwater
pollution. The pollution occurs either when the landfill extends
below the water table or when leachate produced from the landfill
enters the groundwater via an unsaturated zone.

Landfill sites may be located and constructed to accept certain
kinds of waste only or mixtures of different wastes. Domestic

waste is the major source of landfill material and is
characterised by high organic content, metals and glass.
Commercial waste has similar characteristics to domestic waste
with less food waste, more paper, solvents and oil, while
industrial waste can include a very wide variety of materials.

Case history studies have provided some data on the chemical
characteristics of leachates according to the type of waste
tipped, usually expressed in the form of a range of values for
particular determinands. As such the pollution problem is to some
extent predictable though there are many possible variables, in
particular the nature of the unsaturated and saturated zones and
their ability to change and attenuate the leachate plume. Also,
the nature of leachates changes with age and in the case of old
sites there may be no records of the waste tipped.

The range of types of contaminants is inevitably very large.
However, the most common types of contamination associated with
landfill are high salinity (chloride), acidity, BOD, COD, ammonia,
iron, manganese, trace metals and organic compounds.

All water authorities report groundwater pollution problems with
landfill sites to a varying degree, and in many çases it is
regarded as the most significant threat. The problem is most
severe in or near the heavily urbanised and industrialised areas
and on unconfined aquifers. Domestic refuse sites are particularly
important as they represent the major volume of waste. Problems
with very old sites and sites of illegal tipping are also of note.
There are examples of abstraction wells being abandoned because of
pollution from landfill sites. Measurement of pollution from
landfill sites is a significant part of groundwater quality
monitoring carried out by the water authorities.

Most authorities now operate an aquifer protection policy (reference section 4.3) and are satisfied that they are being effectively consulted by the Waste Disposal Authorities. However, although the consultation procedures are regarded as being effective, the number of applications being dealt with is becoming an increasingly onerous workload, and the water authorities may not have the manpower to deal with individual applications as thoroughly as would ideally be required.

In 1982 the Department of the Environment set up a Landfill Practice Review Group to provide guidance on the procedures and practices to be adopted for the continued acceptability of landfill as a means of waste disposal. The findings of this Group have been published as Waste Management Paper No 26 (Department of the Environment, 1986).

The increasing requirement for waste disposal sites is a particular problem in the unconfined aquifer areas of southern England, where it is becoming difficult to find new sites which do not pose a threat of some kind to the environment, including groundwater quality. Given the importance of groundwater resources to public supply the concept of 'dilute and disperse' is being questioned, partly because of the volume of stored groundwater required to reduce pollution to acceptable levels, and partly in view of its feasibility in consideration of recent EC Drinking Water Quality Directives.

3.4.04   Sewage Effluent and Sludge Disposal

The two principal potential sources of contamination associated with the treatment of sewage are those concerned with disposal of effluents and those concerned with disposal of sludges.

Effluents arising from biological treatment plants generally have
a high degree of treatment in terms of removal of BOD and
suspended solids but will contain organic carbon and either
ammonia or nitrate which would cause pollution if they reached the
groundwater. Most effluents from biological sewage works are
discharged to surface water streams and rivers and will not come
into contact with groundwater other than through surface water-
groundwater interaction, examples of which are known.

Some effluents from biological treatment plants are "lost by
absorption" in soakaways and through different types of grass
plots. These systems have a higher potential for pollution of
groundwater than those discharged to surface watercourses. Such
discharges are usually operated and monitored by water authorities
even where the treatment works are operated by private
organisations.

Of perhaps greater concern than treated effluents "lost by
absorption", are the discharges from the vast number of septic
tanks throughout the country. Septic tank effluents do not receive
a high degree of treatment and contain organic matter, ammonia,
viruses and bacteria. Although these discharges receive some
natural treatment as they percolate throught the soil, there is a
chance of contamination of underground resources. Again this is
monitored by pollution control departments within the water
authorities, who can influence the siting and location of septic
tanks.

The greatest threat is probably to private borehole supplies
particularly in volumetrically small aquifers in superficial
deposits. Because of the large number of applications together
with the difficulty of locating unlicensed private wells, the

water authorities do not necessarily have the capacity to consider the applications as thoroughly as they would like.

At present there is active discussion on the need to disinfect sewage effluents before discharge to either surface waters or prior to discharge to land or soakaways. The consequence of such disinfection (especially if chlorine is used) needs careful consideration before it is implemented on a wide scale, as does the practicability of providing effective treatment at the small scale.

The other major potential for contamination arises from the agricultural disposal of sewage sludges, as discussed previously in Section 3.4.02. Those sludges disposed of in solid form would generally release their potential contaminants at a relatively slow rate and would affect surface waters more than groundwaters. More immediate contamination would arise from the widespread disposal of liquid sludge which is the most acceptable from an agricultural point of view (reduced pathogens and readily available nitrogen). However, the high concentration of ammoniacal nitrogen can cause pollution if not correctly controlled.

3.4.05   Accidents, Spillages and Leaks

Accidents and spillages may involve infiltration of chemical substances to groundwater and several such incidents have been reported. Some of these occur on roads and hence have a bearing on road drainage which is discussed in the next section.

The extent of the problem with respect to groundwater contamination depends upon the substance spilled, the potential for attenuation and the type of flow in the aquifer. In addition to incidents involving materials in transit there are examples of pollution being caused by pipeline fractures, fires at industrial plants and leaks beneath chemical storage sites.

Many reported incidents involve complex organic compounds, the behaviour of which is still relatively poorly understood. Clean-up (aquifer renovation) operations may be possible but are frequently difficult and expensive.

Examples of groundwater pollution as a result of accidents are reported by several regional water authorities and locally may give rise to difficult problems. Hydrogeological systems do exist where spillages could result in very rapid transmission of pollutants into water supply, the consequences of which would be extremely serious.

3.4.06   Road Drainage

A significant proportion of the network of major roads in England and Wales is drained by soakaway systems. Where the roads cross unconfined aquifers the possibility occurs of groundwater pollution from runoff and spillages. Perhaps the most vulnerable major aquifer is the Chalk of southern England which is crossed by sections of several major motorways carrying high volumes of traffic.

The extent of the risk is variable. In some cases, adequate liaison exists between the Department of Transport, the emergency services and the water authorities to minimise the risk; for example, drainage can be designed such that groundwater flow is away from abstraction wells. Also, in the event of a spill, the water authority may be informed by the emergency services before substances are washed off the scene of the accident, but under

47

most circumstances this is likely to happen only when a surface
water course is threatened.

There are examples of pollution from road drainage, particularly
increases in chloride following winter salting, and the potential
exists for more troublesome substances to enter groundwater. Some
road drainage systems present particularly high risks; it is for
example reported that the drainage system on the M1-M25
interchange - reputed to carry the largest volume of traffic in
the UK - leads to a series of borehole soakaways, drilled close to
the water table in the Chalk. Since this is in an area where the
Chalk groundwater is used for public supply the risk of pollution
as a result of an accident or runoff is particularly serious, and
it is possible that similar risks exist elsewhere.

The installation of oil traps in the soakaway systems allows
potential contaminants to be pumped out of a lagoon and disposed
of by other means. However, although this system reduces the risk
it does not eliminate it completely as the outlets to the lagoons
have to be manually closed to avoid the spilled liquid entering
the aquifer.

3.4.07   Coal Mine Drainage

Approximately 97% of the 840 Ml/d (1972/73 Figure) of groundwater
which pumped for mine drainage is discharged to waste in rivers
and streams; most discharge occurs in the Yorkshire, South Wales,
Northumberland and Durham Coalfields. Of the total discharge some
58%, according to the 1972/73 NCB survey, came from drainage of
active mines; the remainder came from disused mines which are
dewatered to aid drainage from nearby active workings.

As discussed in Section 3.3.7 the chemistry of groundwater in the Coal Measures is extremely variable. Furthermore considerable and complex chemical changes occur when groundwater enters mine workings. A common chemical reaction involves the oxidation of iron pyrite deposits; soluble salts of iron are formed which can result in a highly ferruginous groundwater. Secondary products include sulphates of aluminium, manganese, calcium and magnesium.

In general the quality of mine drainage water is poor. Most mine drainage waters are acidic and have a total dissolved solids content in excess of 2000 mg/1. The exception is the South Wales coalfield where up to 30% is estimated to be of potable quality.

In terms of quality many discharges from mines are important in maintaining dry weather river flows by providing a constant flow throughout the year. In terms of quality however problems do arise. The River Pollution Survey in 1970 reported that a large percentage of polluted or poor quality rivers were located in coalfield areas and in part this may be attributed to mine drainage sources.

Pollution of both surface water and groundwater have resulted in many specific incidents. For example in Durham, discharge of minewater pumped from disused working caused iron pollution of the River Skerne. Direct discharge of minewater to the Chalk at Tilmanstone, Kent, up until 1974, resulted in chloride concentrations of up to 5000 mg/1 in the Chalk groundwater. Groundwater pumped for supply from the Sherwood Sandstone aquifer in Nottinghamshire has shown contamination due to induction of river water in hydraulic continuity with the aquifer which contains a component of minewater drainage.

Under the Control of Pollution Act, 1974, discharge of minewater is subject to regulation.

3.4.08   Crown Properties

Water supply and waste disposal on Crown Properties is the responsibility of the Property Services Agency (PSA). In terms of groundwater contamination, the main potential sources which are Crown Properties are Ministry of Defence establishments.
As far as legislation is concerned the Crown is exempt from licensing procedures and cannot therefore be prosecuted for failing to comply. However PSA proceed in accordance with legislation in all respects apart from completing applications, making formal agreements or keeping registers, and , in practice, monitor their own supplies and keep their own COPA register.

Close liaison is maintained by PSA with the water authorities in all aspects of water supply and waste disposal including operation of independent abstraction sources and monitoring. In fact, Crown Properties are largely reliant on the water authorities for these services. Where PSA operate their own water supplies, monitoring and treatment is carried out in accordance with 1980 EC Drinking Water Directive (Department of the Environment, 1986) and where such monitoring includes raw ground waters the analyses are usually carried out by the Laboratory of the Government Chemist (LGC).

LGC also routinely carry out scans of raw water quality independently and have identified pollution problems as a result. LGC are currently scanning 20 sites in each of 10 regions per year. Determinations include tri-halomethanes, mercury and other toxic metals, herbicides, pesticides and solvents.

Pollution problems have occurred, particularly at defence establishments in southern England and East Anglia. The events may be long term such as washing of de-icing materials and solvents into groundwater or isolated events such as leaks and spillages; examples of the latter are a burst jet-fuel line and a leaking fuel storage tank. Solvent contamination is widespread on military bases and its origins may go back many years.

Waste disposal is a potential source of contamination in some cases but few problems have been identified. Sewage effluent and solid waste disposal are examples but PSA operate a Technical Instruction on protection of water supplies which minimises problems.

When groundwater pollution incidents occur or are identified and which may be attributable to Crown Properties, the local water authority would be informed. PSA may authorise investigations involving the authority and/or independent consultants to determine the nature and origin of the problem. When the responsibility lies with a Crown Property, the cost of remedial action would probably be met through PSA.

Military establishments are a source of groundwater contamination on a site specific basis and several locally serious incidents have occurred. The time-related aspects of pollution incidents are not well understood and because the activities giving rise to the presence of particular contaminants may have occurred many years ago, the extent of the problem is not clear.

However, there is no evidence that Crown Properties are a greater risk to groundwater quality than many industrial enterprises of a comparable size and nature which are using and storing hazardous materials. However, because of the security aspect and the activities carried out they are regarded as being considerably more sensitive. PSA appear to take an objective view of the problems which arise and liaise effectively with Local Authorities on pollution potential and remedial action, including financial compensation.

3.4.09   Industrial Development and Contaminated Land

Land contaminated as a result of operating or discontinued industrial activity is also a source of groundwater pollution. The concepts involved are similar to those occurring with landfill sites whereby a leachate may be formed which subsequently mixes with groundwater. The range of possible pollutants is very wide, but solvents, other complex hydrocarbons and toxic metals are commonly reported.

Redevelopment of disused sites frequently involves removal of the contaminated soil to purpose built landfill sites, otherwise a possible groundwater pollution problem might result from mobilisation of toxic substances if the soil were only disturbed by reprofiling. Disused gasworks sites are commonly reported as sources of pollution to groundwater. However, problems also occur in modern industry; the incidence of solvents associated with motor vehicle manufacturing in groundwater in the West Midlands is one example. Contamination may also result from leachates produced at the base of mine spoil heaps, though in most cases this will discharge to surface water. Both coal and minerals mining provide examples but the problem is not significant and with respect to groundwater.

3.4.10   Saline Intrusion

Saline instrusion is a potential groundwater pollution problem where unconfined aquifers outcrop at the coast or where old saline groundwaters are present in inland aquifers.

Saline intrusion at the coast has in the past been an important groundwater quality problem. It has occurred where over-abstraction has led to the ingress of sea-water into exploited aquifers and loss of the source due to increased salinity. Examples are the Chalk aquifer in Humberside and adjacent to the River Thames in east London.

The mechanisms involved are well understood as are salinity distributions in most major coastal aquifers. Abstraction in these areas is distributed such that saline intrusion at the coast is either stable or receding, and as such is not an immediate problem nor a major constraint to groundwater utilisation.

Deep saline waters are also known to occur in certain inland aquifers in particular the Permo-Triassic sandstones of northern England and north Wales (reference Section 3.3). The distribution of the saline groundwater and its relationship with the fresh groundwater in the same aquifer has been investigated through research projects and monitoring. Groundwater development, including abstraction patterns and borehole design, has to be carefully controlled to avoid up-coning and lateral movement of the saline groundwater with consequent pollution of the freshwater resources.

3.4.11   Acid Rain

Acid rain is a product of the industrial age and is due to the
emission and dissolution of gases in the atmosphere, chiefly as a
result of fossil fuel burning activities. Sulphur dioxide and
nitrogenous oxides emitted in gaseous form are oxidised and
dissolved in atmospheric water to form sulphuric and nitric acids;
after precipitation as rainfall the ratio of sulphuric to nitric
acid is approximately 2:1.

Acid rain is known throughout the industrial world of North
America and Europe. In Scandinavia for example there are
particular problems of acid rain pollution affecting forests,
rivers and lakes. Rainfall in the UK typically has a pH of 4.2 to
4.5, although pH levels are somewhat higher in the west. There are
no reliable data for long term trends. In 1984 the British
Geological Survey (BGS) undertook a study for the DoE into the
susceptibility of UK groundwaters to acid rain, which was the
first of its kind in this country.

Groundwater contaminated with acid rain may result in lower pH
levels and two possible consequences of this have been identified:

o   corrosion of pipeworks associated with water supply
        distribution systems may be promoted.

o   dissolution of trace metals, occuring either in the aquifer
        source rock or in pipeworks, may result in increased
        concentrations in water supplies which could exceed levels
        set by EC Water Quality Directives.

The BGS report has emphasised the importance of chemical reactions
between rainfall, as it recharges the aquifer, and the soil/rock

mass. The ability of a soil/rock mass to attenuate acidity in recharge water has been termed the Acid Neutralising Capacity (ANC) and this property primarily depends on the amount of calcium carbonate present. The ANC for limestones is thus infinitely higher than that for sandstones, although even in the latter small amounts of calcareous cement can provide a significant neutralising effect. As a result groundwaters in carbonate aquifers rarely have pH levels less than 6.5.

The BGS report includes a map for each regional water authority area showing zones of groundwater susceptibility to acid rain. The zonation is based on knowledge of the geology, particularly the calcareous nature of the rocks, and on the hydrochemistry of groundwater where known. In the latter category, groundwaters with an alkalinity of less than 100 mg/l were considered to represent relatively susceptible aquifers. Susceptibility of the main aquifers in England and Wales may be summarised as follows:

o   The Chalk, Jurassic, Magnesian and Carboniferous aquifers, being essentially composed of carbonate rock, have an almost infinite capacity for acid neutralisation and therefore there is unlikely to be any hazard to groundwater from acid rain.

o   The Permo-Triassic sandstones where a calcareous cement and/or drift cover is present have a low susceptibility and no problems are foreseen. However in shallow groundwaters below outcrop, where there is no drift, and decalcification by weathering has taken place, groundwaters may be susceptible, notably in the north west of England and in Shropshire.

o   Minor aquifers, such as the Millstone Grit, Lower Greensand, Tertiary, superficial deposits and Lower Palaeozoic, which have a low calcareous content and contain low alkalinity groundwaters are considered to be susceptible. Particularly vulnerable are the numerous small, private water supplies obtaining groundwater from shallow wells and springs in susceptible areas, which tend to be widespread over much of Wales, northern and western Britain.

From available data there is no long term downward trend in alkalinity levels which would best indicate the effects of acid rain as presented in the BGS report. However as pointed out in the report, the long term data that are available are from major water supply aquifers which generally have no or low susceptibility to acid rain. The BGS report highlights the fact that for the most susceptible areas there is little or no water quality data by which to judge the effects, if any, of acid rain.

3.4.12   Non Agricultural Sources of Pesticide Pollution

Other organisations, particularly Local Authorities, public utilities and nationalised industries also apply significant quantities of pesticides. Applications of pesticides by private domestic users also occurs. There is no recently collated information on the non agricultural uses of pesticides but it is believed that the principal pesticides are creosote and tar oils for wood preservation together with sodium chlorate and triazines. Within England and Wales, the two latter groups of pesticides are probably applied at a rate of several hundred tonnes of active ingredient each per annum.

Several cases of pollution due to non-agricultural use of pesticides have been reported as well as cases of careless storage and handling, and it is likely that a relationship exists between the two observations.

It is usually difficult to identify the source of pesticides in drinking water because these substances are widely used. However, the water authorities have indicated that, to date, the group of pesticides found most regularly and at the highest concentrations are the triazines. The use of these pesticides in agriculture is relatively limited due to the fairly low area of crops to which triazines can be applied. There is increasing evidence that the occurrence of triazines in drinking water is largely due to non agricultural applications. The main use of these pesticides is believed to be non selective control of weeds on road verges, playing fields, around housing and industrial estates and on railway tracks.

3.4.13   Groundwater-Surface Water Interaction

In most river basin systems there is a degree of interaction between groundwater and surface water. As discussed in section 2.3.3 groundwater may provide a significant component of the total flow in a river as baseflow in summer months. Notable examples are the chalk groundwater fed stream and rivers of Wessex, Hampshire and the Chiltern Hills. Conversely rivers may lose flow to groundwater, given hydraulic continuity between the two and a groundwater level below river level. Such behaviour is common where groundwater pumping stations located near to rivers have lowered the local water table below river bed level.

Both phenomena, groundwater accretion to rivers and river loss to groundwater, have important implications for groundwater quality issues. Examples where problems can and sometimes do arise are as follows:

o   Land Waste Disposal: contaminated leachates from landfills
     may persist sufficiently to affect surface water quality
     where groundwater contributes locally to surface water
     flow.

o   Minewater Discharge: discharge of pumped minewater
     invariably and adversely affects the quality of river water.
     Further downstream groundwater contamination may also occur
     if near-river pumping stations cause induction of poor
     quality river water into aquifers. For example, in
     Nottinghamshire pumping stations utilising the Sherwood
     Sandstone aquifer have been adversely affected by minewater
     contaminated rivers.

o   Sewage Effluent Discharge: most treated sewage effluent is
     discharged directly to streams and rivers. Groundwater
     abstraction can influence overall water quality in two
     ways:

     (i)   where groundwater baseflow is a significant component
           of a river which receives effluent, groundwater
           abstraction can result in less baseflow dilution of the
           effluent and therefore lower river water quality can
           result.

     (ii) in the context of groundwater quality, riverside
           groundwater abstractions may induce inflow of river
           water containing effluent thus causing quality problems
           to water supplies, and examples are known.

3.5 Current State of Groundwater Quality

This section presents the findings of the study with respect to the current state of groundwater quality. The description refers to quality of groundwater currently abstracted or potentially of use based on studies carried out in the course of the project.

Although a number of potentially serious problems have been identified the quality of groundwater used in public supply in England and Wales would appear to be good. We continue to be approximately 30% reliant on groundwater for public supply and there is currently no evidence of this reliance on groundwater diminishing significantly as a result of quality deterioration.

Problems however do exist and, as discussed further in Section 6 are likely to worsen. Nitrates are the major inorganic problem and high nitrate concentrations in some groundwater supplies have necessitated blending and relocation of abstraction wells. Landfills, with the variety of contaminants which they produce, threaten many sources but there are few examples of sources having been lost.

There are a significant number of reported problems with organic pollution, mainly solvents, pesticides and fuels/oils. So far these are point or multi-point pollution problems but the currently available information suggests that as sampling and analytical techniques develop their occurrence will be shown to be more widespread.

4. LEGISLATION AND POLICY ASPECTS

4.1 Introduction

This section contains a review of aspects of legislation and policy which have a bearing on groundwater quality in England and Wales.

The requirements of Acts of Parliament and EC Directives relevant to abstraction and quality of groundwater are outlined, followed by a review of current policy on groundwater protection. The present situation with respect to drinking water quality standards is described separately, followed by an assessment of the implications of the current proposals on privatisation.

4.2 The Legislative Framework

4.2.1 Background

Various types of legislation, regulations and guidelines impinge upon the use and quality of groundwater in England and Wales, and fall into the following categories:

o   Acts of Parliament
o   European Community Directives
o   Industrial Regulations relating to industrial activities
o   Guidelines and codes of practice with respect to industrial and agricultural activities

EC Directives are not enforceable through British law and cannot therefore be regarded strictly as legislation. However, member states are required to take action to comply with directives through the administrative instruments of government. This is an effective system, without being subject to British law. The last category, though not enforceable through law, forms an important part of the overall framework of control.

The manner in which the various categories inter-relate and overlap is complex and is described below by considering groundwater under headings of abstraction, use for drinking water and pollution.

4.2.2   Groundwater Abstraction

The principal legislation relating to abstraction of groundwater is the Water Resources Act 1963 and the Water Act 1973.

Section 24 of the Water Act 1973 sets out the duty of the water authorities to carry out a survey of the water resources within their area, and to undertake periodical reviews of the survey.

Part IV of the Water Resources Act 1963 deals with the general provisions as to abstraction of water, and the procedures for granting of abstraction licences. The objective of this part of the Act is to enable the water authorities to conserve the water resources within their area.

Insofar as groundwater is concerned, there are some exceptions from the general licensing restrictions:

o   single abstractions not exceeding 4500 litres (1000 gallons)

o   a series of abstractions, the aggregate of which does not exceed 4500 litres

o   abstractions by an individual for his/her own domestic
         purposes

o   exploratory abstractions relating to water resource
         investigations

In the context of national groundwater abstraction, these are not
thought to be significant exceptions.

In Part VI of the Water Resources Act 1963, the water authorities
are given powers to compulsorily acquire land in connection with
abstraction schemes, through the Acquisition of Land
(Authorisation Procedure) Act 1946.

The Water Act 1973 saw the dissolution of the Water Resources
Board and river authorities referred to in the Water Resources Act
of 1963. The setting up of the present system of regional water
authorities followed. The water authorities were charged with the
duty of conserving, redistributing or otherwise augmenting water
resources in their area, and with securing the proper use of the
water resources or transferring any resources to the area of
another water authority. These duties are carried out through the
Water Resources Act 1963 and Water Act 1973 legislation.

In essence then, abstraction of water from underground resources
is subject to the provisions of the Water Resources Act 1963. All
abstractions, with the exceptions listed above, must be licensed
by the water authority in whose area the point of abstraction
lies. In making licence determinations, the authority must seek to
conserve water resources. The applicant has a right of appeal to
the Secretary of State should he be dissatisfied with the decision
of the water authority.

4.2.3   Groundwater Use for Drinking Water

The Water Act 1973, Section 11, gives the water authorities a duty
to provide a wholesome supply of water within their area of
supply. The local authority is charged with taking steps to
ascertain the wholesomeness of the water, formerly by reference to
WHO European standards but now by reference to EC Directive
80/778/EEC, relating to Quality of Water intended for Human
Consumption (op.cit.).

EC Directive 80/778/EEC also provides for the use of water in the
food industry, but recognises that regulations covering any
particular industry may require the water to be of a higher
standard than the EC Directive minimum.

Where water authority areas include the area of supply of a
statutory water company, the Water Act 1973, Section 12, requires
that the responsibility for the wholesomeness of the water in the
area be discharged through the water company.

Water quality standards are discussed further in Section 4.4.

4.2.4   Groundwater Pollution

There are several sources of legislation which are concerned with
the pollution of groundwater. Some are specific to groundwater,
others only include passing reference to groundwater. In the UK,
most regional water authorities have strengthened standing
legislation by formulation of their own aquifer protection
policies. Aquifer protection is discussed further in 4.3.

The Control of Pollution Act 1974 Part II is concerned with the
prevention of pollution of water, including underground water.

Control of discharges into underground strata had hitherto been effected through Part VII of the Water Resources Act, 1963. The water authorities are given control over discharges and other activities which could give rise to aquifer pollution. All discharges to the aquatic environment must be given consent by the water authority, except in cases where the authority itself is discharging (at sewage treatment works for instance), where consent must be granted by the Secretary of State. Underground water which is, or may be, used by the authority must be specified by the authority in a document available for inspection by the public.

The Act recognises agricultural practices as a potential major source of water pollution, and Section 51 of the Act gives the authority power to issue a notice to prevent certain practices in sensitive areas. Alternatively, the authority may designate areas over which certain activities could result in pollution of underground waters as aquifer protection zones.

Part I of the Control of Pollution Act 1974 deals with waste on land. It has been recognised in this part of the legislation that waste disposal sites are also a potential major source of pollution. Consequently, Waste Disposal Authorities must consult the relevant water authority when preparing their own waste disposal plans, or before issuing a licence for a private disposal site. Where the two authorities cannot come to agreement, the Secretary of State is empowered to determine the matter. Points of interest here are any conditions relating to site reinstatement and aftercare which may be required by the water authority for protection of underground waters in the longer term. These conditions must be attached to the planning permission for the site, issued under the Town and Country Planning Act 1971, rather than forming part of the licence issued through the Control of

Pollution Act. Generally, the operating system of the Planning Act
outlined by the Town and Country General Development Order 1977,
should, but not by law, include consultation with the water
authority. Ideally, reinstatement and aftercare matters should
form part of the licence issued through the Control of Pollution
Act. However COPA is only concerned with the operation of the
site. It is understood that the DOE are currently considering
strengthening the Act in this area. A further matter of concern is
the high level of technical monitoring and control that is
required for waste disposal sites, as measures to prevent
pollution become more elaborate. Future revisions to COPA should
include statutory provision of these functions.

There are certain developments which are exempt from planning
permission but they are not likely to be of the type which could
result in pollution of underground water. However, sites which do
fall into the latter category should be considered by the water
authority through the provisions of the Control of Pollution Act
Part II.

The EC Directive on the Protection of Groundwater Against
Pollution Caused by Certain Dangerous Substances (80/68/EC)
stemmed from an earlier directive (76/464/EEC) on general aquatic
pollution from these substances. In the DOE circular 4/82
explaining the 80/68/EC Directive (Department of the Environment
1982) it was noted that the UK has adequate legislation to control
discharge of the so-called List I and List II substances, through
the Control of Pollution Act 1974 Part II. There is, however, an
inconsistency between the Directive and the Act. The former
requires that the List I and II substances be prevented from
entering any groundwater whereas the Act refers to specified
underground waters. The result is that the authorities are now

defining larger aquifer protection zones than had been envisaged at the time of the drafting of the Control of Pollution Act.

The disposal of sewage sludge to land is a potential source of pollution of underground waters. Disposal at landfill sites is subject to the provisions of the Town and Country Planning Act 1971 and the Control of Pollution Act 1974 Part I. However, the application of liquid sludge to land for agricultural purposes (reference Section 3.4.02) is apparently exempt from licensing due to the complexities of the ensuing administrative procedures. The practice has been the subject of several issues of guidelines from national bodies over the past 10-15 years. The aspect which has been given prominence in guidelines is the potential for contamination of soil and crops, rather than of underlying groundwater. Nevertheless, the water authorities have all considered the potential threat of pollution of underground water.

EC Directive 86/278/EEC of June 1986 relates specifically to the protection of the environment, particularly soil, when sludge is used in agriculture. In the UK, the water authorities use the DOE/NWC standing Technical Committee Report No 20 as a core document for their own guidelines for disposal of sludge to land. The agricultural community receive guidance from the Ministry of Agriculture, Fisheries and Food through the Agricultural Development and Advisory Service (ADAS). Authority guidelines generally include sections which are designed to protect underground water. At the present time, the water authorities and the DoE are in the process of writing a national code of practice for use of sewage sludge in agriculture. It is important that the final document recognises the threat that increased agricultural use of sewage sludge poses to groundwater quality. Nitrate, bacteriological, and List I and II substance pollution all need

careful consideration. Also, procedures for monitoring and control
of the practice in relation to groundwater quality must be
tackled.

Although the use of sewage sludge in agriculture is a potential
source of groundwater pollution, there appears to be little
documentary evidence to support the theory. This may be because
little or no research has been directed this way, or indeed that
the danger is overstated. In either case, the likely increase in
the practice will require strengthened guidelines, based upon
research into the potential hazards.

The Food and Environmental Protection Act 1985 includes provisions
for protection of the environment from contaminated foods and from
pesticides, which are widely defined to include agrochemicals,
wood preservatives, industrial and amenity herbicides,
rodenticides and surface coatings. Implementing regulations on the
control of pesticides places a general obligation on all those who
supply, sell, store and use pesticides to safeguard the
environment and a particular obligation on users to avoid
pollution of water; guidance on meeting these obligations is
provided in Codes of Practice which will become statutory codes.
The regulations also require anyone carrying out aerial spraying
of pesticides to consult the local water authority if the area of
spraying is adjacent to water. The British Agrochemical Standards
Inspections Scheme is a voluntary registration scheme aimed at the
Act, and it involves consultation with water authorities.

Further potential sources of groundwater pollution are industrial
accidents. The Control of Industrial Major Accident Hazards
Regulations legislate the preparation of emergency plans to deal
with such accidents. The procedures necessarily involve

consultation with water authorities where groundwater quality could be threatened.

4.3   Aquifer Protection

The necessity for policy on protection of underground water is widely recognised and is an issue which has been the subject of a considerable amount of discussion and debate in recent years.

Attempts to protect groundwater in England and Wales are covered by statutory requirements and less formal policies adopted by the relevant authorities as described in Section 4.2.

Under the COPA I, COPA II and various EEC Directives the water authorities are formally involved in the decision making process while under other legislation they may be consulted only when the planning authority consider that the issue is relevant. It is partly in response to these requirements that several water authorities have formulated an Aquifer Protection Policy (APP) to provide a framework for objective decision making in the planning and consultation process. The more comprehensive APP's have proved extremely useful by providing consistent standards of approach and avoiding conflicts.

It is important here to distinguish between the concepts of aquifer protection zones and an aquifer protection policy. An aquifer protection zone refers to an area specified by a water authority as part of COPA II (reference Section 4.2) and is usually an area around an abstraction source. Aquifer protection policies refer to measures to control pollution over a whole catchment or aquifer, as described below.

The advantage of a comprehensive and technically sound APP goes beyond the legal requirements providing as it does a sensible code of practice which encourages prevention of pollution. Aquifer vulnerability can be established through hydrogeological assessment with the overall aim of locating activities of highest risk in areas of lowest vulnerability. An immediate advantage is that clearly acceptable and clearly unacceptable proposals can be quickly identified, making the decision making process more efficient.

At present some water authorities adopt APP's which meet all these requirements, others have produced broad guidelines relating to source protection and/or waste disposal, while others do not operate an APP as such. In the latter cases, reliance is placed on studies of the particular cases in relation to the appropriate legislation. Authorities with the more comprehensive APP's also accept the necessity in many cases for more detailed study of individual proposals.

Perhaps the most important contribution made by an authoritative APP is that the potential for and management of groundwater pollution is seen in the wider hydrogeological context such that the regional impact can be assessed. That is, the protection measures are aquifer rather than source-only orientated.

However, it is important to note that APP's do not cope well with problems of diffuse pollution and it is difficult at present to envisage, for example, certain agricultural activities being prohibited over wide areas of aquifer outcrop due to perceived risks of pollution from, say, nitrate or pesticides.

4.4   Water Quality Standards

Prior to 1980 the WHO European Standards for Drinking Water (1970) and the DHSS report on Public Health and Medical Subjects No 71 "The Bacteriological Examination of Water Supplies" were accepted as giving guidance on the criteria for the wholesomeness of water supplies. In July 1980 an EC Council Directive 80/778/EEC relating to the Quality of Water Intended for Human Consumption was adopted and implemented in July 1985 by administrative means. Member states were allowed two years to bring into force any necessary legislation and to establish procedures and a further three years to ensure that waters comply with the provisions of the Directive. Waters covered by the Directive are those for consumption supplied by a statutory water undertakers and water supplied by non-statutory undertakers including private suppliers.

The Directive lists for most parameters both a guide level (GL) and a maximum admissible concentration (MAC). Interpretation of GL and MAC values are discussed in the joint circulars from the DoE and the Welsh Office (20/82 (DoE) - 33/82 (WO and (25/84 (DoE) - 51/84 (WO)) which also discuss requirements for monitoring and analyses. In the UK, monitoring of public water supplies is the responsibility of the relevant statutory water undertaker whereas it is the responsibility of the relevant local authority to ensure the application of the provisions to privately owned supplies. Article 12 of the Directive require Member States to take all necessary steps to ensure regular monitoring of water for human consumption and standard sampling for pollution and minimum frequency of sampling and analyses.

It appears to be generally accepted that "water for human consumption" is that entering or contained within the distribution system and the statutory undertakers and local authorities only

monitor this water. However, the Directive standards apply
strictly at the point that supply is made available to the
consumer and member states are to prescribe the monitoring points.
There is not a requirement in the Directive to monitor or analyse
groundwater sources directly.

4.5   The Implications of Privatisation

4.5.1   Background

Privatisation of the water authorities is on the horizon and is
expected to commence within the term of the present parliament. In
July 1987 the government published a consultation paper entitled
'The National Rivers Authority - A Public Regulatory Body in a
Privatised Water Industry'. This was followed in December 1987 by
a further document 'The National Rivers Authority - The
Governments Policy of a Public Regulatory Body in a Privatised
Water Industry', published by DOE, MAFF and the Welsh office. In
the broadest terms, the intention is to set up a National Rivers
Authority (NRA) to take over responsibilities for water
conservation policies and to ensure strict safeguards against
water pollution. The water supply and sewerage functions are to be
transferred to the private sector, in the form of Utility Public
Limited Companies (Utilities). The utilities will be subject to
economic control by the office of the Director-General of Water
Services (OFWAT).

4.5.2   Implications for Groundwater

The NRA will take over from the present water authorities the
powers of control of abstraction from underground strata, through
the legislation of the Water Resources Act 1963. We envisage that,
for each utility area, the NRA will issue a water resource

development envelope, within which the utility must operate. Decisions on which resources are to be exploited, and on the extent of conjunctive use of resources, should be left to the utility. Water requirements of users other than the utilities will be administrated by the NRA who will be required by statute to consult with utilities and water companies about abstraction applications . The national planning of water resources will naturally fall to the NRA, who we believe will recommend and administrate inter-utility transfers where necessary.

Quality of groundwater for use as drinking water will continue to be the responsibility of the utilities through EC Directive 80/778/EC, Quality of Water Intended for Human Consumption.

Environmental protection and pollution prevention will be important functions of both the NRA and Her Majesty's Inspectorate of Pollution (HMIP). Determination of quality objectives and standards to which waters are maintained and improved will be the responsibility of Ministers. However, the assistance and advice of NRA will be an important role. Further, the NRA will have responsibility for achieving the objectives and standards through the issue of discharge consents. HMIP will continue with its role as an advisory body to the Secretary of State, concerning disposal of effluent from industrial processes.

The consultation paper suggests that waste disposal authorities will be required to consult both the NRA and the utilities over waste disposal plans and issue of licences. This could lead to conflicting advice being given and, in its response to the green paper, the Institution of Water and Environmental Management (IWEM) have recommended that the NRA should coordinate the views of all interested parties.

It is expected that the Secretary of State will have powers to designate aquifer protection zones, normally acting on the advice of the National Rivers Authority. Such zones may be relevant to the National Rivers Authority in consideration of applications for discharges to underground strata, and when considering the possible effects of agricultural practices. They may also be relevant to the development of sludge disposal policies, a matter primarily for the utility companies.

An area in which the effects of privatisation will be seen is the regional contracting-out of some of the NRA's functions. In particular, the routine sampling and analysis of rivers, groundwater and effluent discharges is thought to be suitable for letting to competitive tender. The utilities will be in a strong position to offer such services but there are other commercial organisations who can also offer these services. In view of the many problems associated with groundwater sampling described in 3.2.3, tender documents for the work will require very detailed specifications. It may be that groundwater sampling and analysis should be contracted-out separately from surface water sampling, as it is likely that the number of potential contractors will be limited.

We envisage that the present statutory duty of the water authorities to undertake research will be retained by the NRA. However, the utilities will also be motivated to undertake research into engineering and process matters by their need to meet standards and improve efficiency. It is likely that the NRA and HMIP will take on the responsibility for long term national research, whilst the utilities are more likely to be interested in research which can offer gains in the shorter term.

5.   RESEARCH

5.1   Introduction

5.1.1   Funding Bodies

Research into UK groundwater quality is funded principally by the
Department of the Environment, the Natural Environment Research
Council and the water authorities. Research impacting upon
groundwater quality is also funded by the Ministry of Agriculture,
Fisheries and Food, and the Agricultural Development Advisory
Service (ADAS). Increasingly, funding is being sought from the
European Community. Apart from research into the effects of
specific products marketed by industry, and some landfill related
research, funding by industry in the groundwater quality field is
extremely limited, although in the specialist quality aspects of
radioactive waste disposal NIREX provides funding.

5.1.2   Research Organisation

The research is currently being carried out by the Water Research
Centre, British Geological Survey, the regional water authorities,
various Universities and specialist private consulting
organisations. Allied research is carried out by agricultural
groups such as ADAS. Because of the intimate inter-relationships
between surface and groundwater, research being undertaken by the
Institute of Hydrology, the Freshwater Biological Association and
through the Nature Conservancy Council also forms a part of the
current programme.

5.1.3   Extramural Funding Procedures

The procedures for distributing research funds differ between the
various funding bodies. The Department of the Environment operates
a sensible dual system whereby once the principal fields of
national interest are identified by the Department tender bids are
invited. Unsolicited proposals are also considered and funded when
thought pertinent to the overall programme. A similar procedure is
adopted by NIREX.

Funding through the Natural Environment Research Council is by
grant or research studentship application, exclusively through
tertiary education establishments and research institutions.
Awards are determined by peer review. The Council encourages
industrial participation and funding support in the research.

The water authorities carry out much of their own research but
also operate through tendering procedures with consultants. For
specific works, specialist consultants may be identified under
single tender arrangements. As the water authorities provide
funds to support the Water Research Centre much of the groundwater
quality research has traditionally been carried out by the Centre
which has good liaison with the authorities.

5.1.4   Co-ordination of Research

The overall structure of British Government funding does not
permit a fully co-ordinated groundwater quality research
programme. The Department of the Environment and the water
authorities are normally reasonably aware of each others
activities and have some collaborative research. For major quality
investigation programmes, as for example with nitrate, the
Department may convene a wide ranging consultative group. Such a

practice is of considerable value for co-ordinating research and also assists the Department in identifying new potential research areas.

Because the Department has national responsibilities in groundwater quality, it is subject to very short-term government directives as well as long-term strategic research requirements. The need for ad hoc short-term research to meet unforeseen circumstances is inevitable although perhaps with a more co-ordinated national awareness of potential pollution problems it could be reduced. The short-term requirements clearly relate to the longer-term strategic research programme.

In 1986 the Department of the Environment published 'Water Research in the Longer Term' (Department of the Environment, 1986) which sets out the findings of the Long Term Water Research Requirements Committee. With respect to groundwater it was recommended that further investigation of the causes and effects of contamination of groundwater should be given high priority.

In certain fields strategic groundwater research may require a very long time span, and extend beyond the life of a Parliament; for example, UK nitrate contaminant research has been running for some 10 years. As a result, strategic groundwater research may be vulnerable to political changes as has been the case with radioactive waste disposal research. Again, political influences on research planning are inevitable although there should be more awareness in the Government that a consistent, well funded, long-term research policy is essential. The curtailment of the shallow disposal of low level radioactive waste investigation programme in 1987, is an example of a change in political policy which has been particularly detrimental to groundwater quality research far beyond the particularities of radionuclied transport.

Between water authorities a good exchange of research information on established major topics takes place. An effective forum for the exchanges of potential or imprecisely defined groundwater quality problems does not exist but would be of value for identifying future research requirements.

The Natural Environment Research Council tends to consider research on a wider and more fundamental basis than the Department of the Environment and the water authorities. The Council is generally aware of the other research programmes. The peer review procedure enhances the Council's awareness of other research programmes and this is currently being extended through the convening of consultative expert review committees.

In summary, the UK research funding bodies tend to be aware of each others research programmes indirectly, through consultative groups or committee participation, registers and publications etc, but no overall co-ordination of research or research requirements exists. While it is appreciated that within the government strictures a co-ordinated policy may be difficult, more effort at co-ordination through a liaison group is desirable, particularly in groundwater quality research which is progressively becoming more multi-disciplinary.

Increasingly, research funding opportunities are occurring through the European Community. The grant awards can be much larger than those available from the national funding bodies. The general impression, however, is that many British research organisations do not take full advantage of the opportunities and there is a lack of awareness of the Commission's call for tender. The problem is, to some extent in the Commission bureaucracy and the difficulty of easily obtaining tender call information in adequate time to submit tenders. In addition to the problems of obtaining the initial information, the follow-up of tenders through British Government departments can be extremely difficult, and a freer liason between research organisations and the DoE would be advantageous.

Some organisations such as the Natural Environment Research Council, which offers services at a fee to universities etc, maintain a worthwhile presence in Brussels. Nevertheless it would appear that the British do not perform adequately in obtaining research funds in comparison to their continental counterparts.

As the EC can provide groundwater quality research funding in addition to national funds it would be of advantage to all concerned if the Department of the Environment could provide a more effective means of helping research applicants than is currently the case. Identifying research areas that would be supported by British representatives on Commission committees, helping identify other European research organisations for joint applications and processing tender queries through to the Commission, are some of the aspects that should be covered. As such services are inevitably constrained by staff availability it would be advisable for the situation to be reviewed internally by the Department. Nevertheless, a much more effective system to assist in obtaining research funding is definitely required.

5.2   Research Expenditure

In Table 5 details are given of the research expenditure for the broad areas of groundwater quality research pursued over the period 1982-87.

TABLE 5

EXPENDITURE (IN £1000'S) IN FINANCIAL YEAR SHOWN

ORGANISATION	1982 - 1983		1983 - 1984		1984 - 1985		1985 - 1986		1986 - 1987	
	Nitrates Research	Non-Nitrates Research	Nitrates Research	Non-Nitrates Research	Nitrates Research	Non-Nitrates Research	Nitrates Research	Non-Nitrates Research	Nitrates Research	Non-Nitrates Research
(i) Research Institutions										
BGS	Total during period 1982 - 1987: Nitrates, 453, Non-Nitrates 958									
IOH	Unable to provide data exclusive to groundwater quality									
WRc	100	280							30	386
DoE (Research Spending)	80	125	121	142	111	105	123	150	143	136
(ii) Water Authorities										
Anglian	18	0	72	5	62	0	54	77	98	36
Southern	4	0	35	4	5	4	3	4	8	4
Wessex	0	0	0	0	0	2	0	2	15	0
South West	2	0	0	0	0	0	0	0	0	0
Severn-Trent	0	26	0	4	0	5	12	24	0	46
Thames	0	25	0	10	0	12	0	75	0	85
Welsh	Generally low expenditure on organics and non-nitrate inorganics; no nitrate research									
North West	Data available: 1977 - 1984 100 on nitrate, 1978 - 84 250 on salinity studies									
Northumbrian	0	3	0	3	0	3	0	3	0	3
Yorkshire	Total approximately 65 on nitrate/non-nitrate research combined									

RESEARCH EXPENDITURE ON GROUNDWATER QUALITY

The data provided do not present a clear statement on trends in levels of funding. Many are estimates only due to the difficulty faced by some organisations in separating groundwater quality research costs from other water quality research. Within the water authorities highest groundwater quality research expenditure is incurred by authorities whose reliance on groundwater is heaviest.

5.3   Current Research

The various areas of groundwater quality research for this Section of the Report and for Section 5.4 are considered under the groupings of general inorganic, nitrates, organic and landfill. It is appreciated that the groupings are not exclusive but are used for convenience of discussion.

5.3.1   General Inorganic Chemical Quality Research

The general natural inorganic quality of groundwater in the UK is particulary well documented through extensive research and monitoring programmes. Research concerning the major inorganic constituents has resulted in a reasonable understanding of the hydrochemical processes that influence their evolution, although it could undoubtedly be improved. Numerical modelling techniques have kept pace with the understanding of processes.

For the major aquifers that provide the bulk of public supplies and much of the industrial supplies, base-line natural major inorganic constituent concentration distributions are known sufficiently to allow contamination to be identified, although they are not nationally collated.

Long-term research programmes, and sensible management policies by the water authorities, have minimised the threat of induced contamination from naturally occurring brackish or saline groundwaters juxtaposed to fresh groundwaters. Modelling of the inter-relationships has been adequately covered in groundwater resources terms although research is proceeding into long-term monitoring techniques.

Under the auspicies of the Department of the Environment, the UK Acid Waters Review Group has been effective in determining the broad influences of acid rain on groundwater quality, indicating that generally few problems exist in the major aquifers, although specific research requirements do exist and are considered below.

Currently research is underway in determining trace constituent concentration and process controls in groundwaters. The ensuing data will provide a valuable constituent base-line for certain aquifers. Research interests in trace metals in recent years have included natural iron occurence in groundwaters, metal concentrations in peat influenced groundwaters and metal contaminants from animal wastes.

Long standing research into isotope concentrations in groundwaters has been of value in the UK indirectly with respect to groundwater quality by contributing to the understanding of groundwater travel times. Isotopic research work has been enhanced over recent years by the significant funding of radioactive waste related studies. These studies, which are ongoing, have contributed substantially to the knowledge of transport phenomena, sorption etc and have allowed the development of sophisticated hydrochemical-hydraulic models. Of major interest has been the development of risk assessment techniques. Both the modelling and risk assessment work should prove of considerable value in other groundwater contaminant fields.

5.3.2   Nitrate Research

Over the past decade nitrate research has been the major groundwater quality topic. As a result the UK has made significant advances in forecasting some of the dominant process controls. The success of the nitrate research can in part be attributed to the sensible co-ordination policy fostered by the Department of the Environment through a Nitrate Sub-Group and the Nitrate Co-ordination Group. Liaison between the groundwater and agricultural research organisations, however, has taken some time to establish.

A substantial amount of research has been carried out on nitrate movement in the unsaturated and saturated zones, particularly related to non-point source contamination, although further research is required as discussed below. A very good standard of modelling of nitrate transport has been achieved for distributed and site specific forecasting.

Research is currently underway to examine microbiological effects on nitrate movement in the unsaturated zone and also to examine nitrate concentrations in aquifers beneath differing land use areas.

As the understanding of nitrate transport has developed so has the awareness of a need for aquifer protection policies. Such policies have been adopted by some water authorities and a start has been made on the preparation of maps showing inter-relationships between soils, geology and land use which should assist in assessing aquifer vulnerability to nitrate contamination. Encouragingly this work is resulting in more co-operation between groundwater and agricultural interests.

As an alternative, or in addition to the protection of groundwater from nitrate contamination, the removal of nitrate from groundwater is possible and in recent years has been extensively researched. Most emphasis has been placed upon post-abstraction treatment although experimentation with in-aquifer induced denitrification has been carried out and is ongoing. Research has also been carried out to examine the relative cost implications of treatment, protection and possible strictures on agricultural practice.

5.3.3 Organic Chemical Quality Research

Research in the UK into organic concentrations and the control processes has only recently commenced in any serious form. Two main fields of research reflecting contamination concerns have been identified, these are pesticides and industrial solvents. Efforts are being particularly concentrated as a result of the implementation of the EC directives.

A range of research initiatives are underway with examinations being undertaken of sampling procedures and laboratory analysis techniques. Water authorities are commencing sampling and analysis programmes to establish, where possible, organic concentration distributions.

Specific research projects are concerned with known pesticide pollution occurrences, degreasing agent pollution beneath industrialised areas (chiefly trichloroethylene) and fuel spillages. In the two former cases, the main preoccupations are with sampling, analysis, concentration distribution and in the case of trichloroethylene, understanding something of the water immiscible - miscible characteristics.

Fuel spillage research has been active longer than most other
organic contaminant research. Several spillages of different types
of fuel have been sampled and a degree of understanding of
attenuation processes such as volatilisation and sorption
obtained. Some numerical modelling of fuel spillages evolution is
underway.

5.3.4 Landfill Research

Landfill research has been extensively funded by the Department of
the Environment and a large number of different organisations have
participated.

The general principle of containment for toxic wastes has resulted
in most of the groundwater quality research being directed towards
domestic landfills operating under natural ground attenuation
principles. The adoption of dilute and disperse landfill sites has
been leniently viewed by many water authorities in the past with
possible leachate contamination assessed through the unsaturated
and saturated zones in relation chiefly to chloride and BOD
concentrations.

The Department of the Environment has issued technical memoranda
on waste management, which are generally followed through the
industry, and which have been based upon the Department's research
programme. Through the 1980's research interests in landfills has
been directed chiefly at management procedures, pre-treatment,
re-cycling, gas generation and control etc. Nevertheless, research
related to leachate generation and its effects on groundwater
quality has continued through from its heyday in the 70's with
studies of capping materials, liners and site specific plume
development. Current research also includes studies of the

mobility and attentuation of organic pollutants and the modelling of their migration.

Research has been carried out by industry but many of the results are not fully available. There exists, however, a wealth of data about landfill sites that can be built on for further research which will undoubtedly be necessary with the implementation of the rigorous EC water quality directives.

5.4   Research Requirements

5.4.1   Introduction

The past UK research achievements in groundwater quality have been noteworthy, such as in the field of nitrate, so that a good foundation of expertise exists on which a future programme can be built. As discussed below, however, the threat to groundwaters is increasing, particularly from organic pollution and a more multi-disciplinary research approach, with more co-ordination between research establishments than in the past, is necessary.

In the following discussion of the research requirements and for any funding, it is essential that the fields of research are viewed within the overall hydrogeological context in that the studies of specific determinands must not be divorced from co-existing hydrochemical, matrix and hydraulic influences.

For future research it is stressed that the main thrust should be towards proper sampling techniques, advanced analytical instrumentation, the elucidation of hydrochemical processes and the development of containment management techniques.

Groundwater hydraulic and hydrochemical modelling both in
deterministic and statistical terms, has largely overtaken the
ability to obtain, analyse and interpret hydrochemical data so
that while modelling forms an integral part of groundwater quality
research, the general view in the industry is that present
modelling techniques are adequate for most aspects in the
forseeable future.

5.4.2   General Inorganic Chemical Research

One of the principle difficulties in determining groundwater
quality is the fact that sampling of the water disturbs its
equilibrium with the system. Any analyses obtained are therefore
not necessarily representative of the sampling location.
Non-representation of analyses is a well known problem in the
industry and is more prevalent in water resources monitoring and
investigations than for example in radioactive waste studies
because of the financial constraints and lack of sophistication of
many of the monitoring wells. An examination of the degree of
non-representation in various hydrogeological systems is long
overdue and would be best performed using major ion chemistry.

Increasingly sophisticated multi-port monitoring systems are
coming onto the market from overseas and while the costs are
decreasing, they are still expensive for comprehensive monitoring
programmes. More initiative is required to develop a UK system,
hopefully within an acceptable price range.

One means of reducing non-representation of samples has been seen
in the UK developments in fluid logging and remote sampling. In
the deep-sea drilling programmes the logging is being extended to
include chemical parameter profiling including pH and redox
measurements, which could be used to great advantage in

groundwater quality studies. The development of such logging equipment for the water industry would be of distinct value.

The need for groundwater quality research in the field of major inorganic constituents in the major aquifer systems is minimal although the collation of base-line data for such items as fluoride, bicarbonate and sulphate would be useful as part of epidemiological research.

In parts of the United States brackish quality groundwaters are protected against pollution for possible future use. With the gradual quality deterioration of fresh UK groundwaters, the potential of brackish waters should be examined in terms of quality, quantity and desalination, options and costs.

In most of the major aquifer outcrop areas, buffering capacities rapidly curtail acid rain effects, nevertheless, local problems do occur. For example significant sulphate concentrations in the Lincolnshire limestone are possibly due to acid rain while the lack of buffering in sandstone aquifers in south-west England could lead to problems. Although of apparently limited significance in the major aquifers, acid rain effects causing increases in undesirable solute concentrations should not be ignored.

While acid rain may not extensively affect our major aquifers the position may be different in shallow aquifers which, although not used for public supply or major industries, are used for local supply purposes. Upland surface waters in places like Wales are showing the effects of acid rain with the mobilisation of aluminium etc and the question is whether shallow aquifers are being significantly contaminated. In all probability other forms of pollution are more prevalent as discussed below.

To further base-line data aquisition the current research into
inorganic trace constituents in groundwaters should continue.
Included in this work should be more determinations of iron
species occurrence and controls; because of increases in nitrates
beneath outcrop areas, abstractions are progressively being moved
to confined aquifer areas where iron contamination is sporadically
present.

5.4.3   Nitrate Research

Much has been achieved in understanding nitrate movement in
aquifers although there are still gaps in knowledge in the soil or
shallow unsaturated zones.

For future research more needs to be known about nitrogen
reactions in the soil zone as this will relate surface nitrate
loading to leaching. Such reactions will partly include
microbiological effects which are poorly understood and which need
to be researched more fully throughout the unsaturated zone.

In the saturated zone two main research aspects require attention,
firstly, whether abstraction of good quality groundwater from
beneath the confining cover down gradient of nitrate contaminated
water results in denitrification in the nitrate waters as they
move into less oxygenated parts of an aquifer, and secondly,
whether more can be understood about the controls on nitrate
concentration layering which is evident in several aquifers and
results in significant seasonal fluctuations of nitrates in well
supplies.

Perhaps the most important aspects of nitrate research, however,
should concern land use practices and their possible changes;
collaboration with agricultural organisations will be essential in

such work. As it is generally accepted that most nitrate contamination of groundwaters occurs from agricultural sources the possibility of changes in land use could affect diffuse nitrate contamination. Whereas leaching rates are reasonably well known beneath arable crops there is very limited knowledge about nitrate leaching beneath grassland under its various forms of management and this should be fully researched. Also current research is showing that nitrate leaching cycles and rates beneath woodlands are not fully understood and require quantification. Allied to this is the need for a broader assessment of groundwater quality effects beneath woodlands in that other ions (notably chloride) have been shown to increase relatively.

As part of the land use studies the research into protection zone techniques for groundwater abstraction sites should be continued. This should embrace much of the existing research methodology but will require detailed monitoring procedures to be established.

Undoubtedly some research will continue into post-abstraction treatment and it would be sensible to further experiment with in-aquifer artifical denitrification. The relative costs for treatment as against protection, and use variation etc. could eventually be critical, and this aspect needs to be more widely examined.

While diffuse nitrate contamination is the main preoccupation, occasionally difficulties do arise with point source nitrate loading from silage storage, liquor lagoons, and feedlots. More needs to be understood about such contamination occurrences through case studies and it would be valuable for guidelines to be produced vis a vis their location in respect to groundwater supply sources etc.

Increasingly, as would be expected, data are showing significant nitrate pollution of private shallow aquifer sources, although indeed nitrates may eventually prove to be only one of many pollutants in such sources. A more comprehensive data base of pollution occurrence, not only of nitrate, would appear to be necessary.

5.4.4   Organic Chemical Research

This is very much an open field in the UK where research is lagging behind some parts of continental Europe and North America. The research requirements bear similarities to those found in the nitrate programme but the organic concentrations are generally one or more orders of magnitude lower. Care must therefore be exercised to avoid funding research that could founder because sampling and analytical techniques may not provide the necessary precision.

Organic research should be the main groundwater quality concern over the next decade and will need to commence with the establishment of reliable sampling techniques. This will mean the testing of suitable materials that will not contaminate samples and the development of sampling systems of the multi-port variety noted in Section 5.4.2. Inevitably organic contaminant investigations will prove expensive so that, from the start, research into the use of portable spectrometers and other site chemical measuring systems would be advisable.

Whereas sampling poses considerable difficulties there are equally problems with laboratory analytical procedures. Many of the toxicity limits set for organics are above or close to the current limits of detection. Research is therefore vital in improving laboratory techniques. Enquires through the industry indicate that there will be a need to establish more organics research laboratories nationally.

In order to start assessing the extent of the organic contaminant problems, the water industry has commenced data collection which inevitably is as yet uneven throughout the country. Emphasis should be placed upon base-line surveys on a comprehensive level as in any case this will effectively ensue from EC requirements.

As part of the survey work attention should be paid to likely source areas such as general agriculture and transportation (motorways and railways) where herbicide applications may be large, airfields where de-icing chemicals and fuel spillages can cause contamination, and industrial complexes from which a large range of contaminants may arise. Unfortunately little is known of organic contaminant migration rates so that although sensible chemical application methods and storage may currently be practised by industry, past uncontrolled practices may have contributed to present day groundwater contamination.

Much of the organic research will need to be directed at process controls. From the current research it would appear that attention should be concentrated initially upon herbicides such as atrazine, organic solvents and fuels, although undoubtedly other organic contaminants will be identified and require researching.

It is considered that the main attenuation controls occur in the shallow unsaturated zone so that representative field studies for differing aquifer-soil types should be targeted to this zone. The studies will require a strong backing with laboratory experimentation in which it would be hoped that the chemical industry could provide expertise. Certainly collaboration with the chemical industry would be beneficial technically and if possible financially.

Deeper parts of the unsaturated zone and the saturated zone should clearly not be excluded from research, although it may be more difficult to identify processes if the main attenuation occurs at shallower levels and contaminant concentrations are not large. Hydrodynamic dispersion and diffusion studies will be important and as with the shallow zone studies will need to be carried out at representative field sites in fissured and non-fissured aquifers with laboratory experimental backing. The miscible - immiscible characteristics of solvents will be an important research field.

As discussed above, a number of hydrocarbon fuel spillages have been examined so that some knowledge of the scale of effects in certain hydrogeological environments has been obtained. With the incidence of spillages likely to increase in the future, an appraisal of existing spillage data would be an advantage so that guidelines could be prepared for the water authorities in order that they may assess the operations that may be required when future spillages arise.

As part of organic pollution studies the renovation of polluted aquifers will undoubtedly require research.

Attention has been drawn above to the possibly serious contamination of shallow private groundwater supplies. While heavy metals and nitrates may cause problems, it is likely that organic pollutants pose the greatest threat and clearly more needs to be known about the degree of this threat.

5.4.5   Landfill Research

The current principles of dilute and disperse for domestic
landfill sites will require more stringent control as a result of
the new EC quality directives with their rigorous organic
contaminant strictures. Whereas assessments have been made of
likely organic contaminants arising from domestic wastes in the UK
and some current research is underway into organic migration below
landfills, the overall knowledge and appraisal techniques, as with
organics generally, are limited.

Department of the Environment guidelines for landfill selection
exist but there is an undoubted need to upgrade these guidelines
so that advanced scientific procedures can be established for site
assessment. Whereas transport and hydrochemical models applicable
to landfills exist, many of the parameters required for the
models, even related to inorganics, are not known in UK
conditions. As a start to upgrading landfill assessment it would
be worthwhile collating existing data for diffusion coefficients
etc and illustrating long-term case histories of monitored sites.
For example, although unsaturated zones are recommended beneath
landfills it is difficult to know what thickness should be adopted
within a particular hydrogeological situation. More investigation
of existing representative landfill sites are necessary to
increase case history data.

The water authorities correctly place the onus on licensees to
show that landfill sites are feasible. The investigations and
assessment costs to the waste industry are increasing but unless
that industry is prepared to spend more on advanced scientific
techniques the future of landfill operations may become seriously
limited.

5.5   Future Research Funding

On a national basis it should be stressed that there is
undoubtedly a need for more funding of research into groundwater
pollution. We consider that the Department of the Environment is
the obvious vehicle through which the funding should be directed,
though a wider number of organisations than are at present being
funded ought to be considered.

6.   FUTURE TRENDS

6.1   Use of Groundwater

Data presented in Section 2 indicate that after many years of
increased development of groundwater resources, its overall
utilisation has stabilised in terms of total licensed abstraction.
The situation in terms of actual abstraction is not completely
clear due to lack of information but it would appear that major
resources are at least committed through licensing and there is no
prospect for significant increases.

On the basis of resource availability, that is, abstraction being
kept to a level which does not exceed the constraints imposed by
recharge replenishment or available storage, it is unlikely that
groundwater utilisation can increase significantly. Equally, there
are no major problems of over-abstraction which would require
groundwater utilisation to diminish, though re-distribution of
abstraction may be required in some areas, for example, due to
problems of low river baseflows.

Changes in utilisation may be brought about by water quality
constraints, as discussed in the next section. Deterioration in
quality or changes in required quality standards could both lead
to treatment requirements beyond practical or economic levels,
with a consequent reduction in groundwater utilisation.

6.2 Quality of Groundwater

6.2.1 Relevant Changes in Agricultural Activity

a) Land Use Changes

Significant changes in land use have taken place in England and Wales over the last 20 years. During this period the total area of grassland has declined by 14 per cent, most of which was due to temporary grass being converted to arable cropping. During the same period the area of cereals has shown an overall increase of 6 per cent. Most of this increase has occurred over the last decade as cereals have replaced the declining area of root and horticultural crops. Over the last 10 years there has been a dramatic increase in the area of oilseed rape grown.

Currently, there is a considerable variation in the distribution of crops and grass throughout England and Wales. Arable cropping predominates in the drier eastern parts of the country while grassland dominates in the west.

Future changes in land use will have a major impact on the distribution of the nitrate and pesticide pollution load. At present a great deal of uncertainty faces the farming industry and it is extremely difficult to predict the outcome of negotiations between Member States of the EC. Whatever the result, economic and political influences are unlikely to be uniform throughout England and Wales. Likely future trends in land use are summarised below:

i)   Uplands

Income support will probably continue; consequently apart from
further expansion of policy little change in agricultural land use
is likely to occur and any change will be relatively slow.

ii)   Productive Lowlands

Farming will continue to be the major rural industry in these
areas. Again, relatively little change in land use is anticipated
unless extensification programmes, such as 'set aside', are
introduced.

iii) Marginal Lowlands

These areas will probably experience the most dramatic change in
land use. Marginal lowlands are likely to include thinly soiled
areas over chalk downland and other limestones, light drought –
prone land without irrigation, river valleys, urban fringes, heavy
wet lowlands and sloping arable land subject to erosion.
Agriculture within these areas is likely to become less intensive
over the next 10 years, and reversion to grassland or
establishment of woodland is likely.

Groundwater quality may therefore benefit from this reduced
agricultural intensity particularly in aquifers under parts of the
chalk downland and on areas of light sandy soils without
irrigation. However, the impact on groundwater quality may not be
felt for several decades.

b)   Inorganic Nitrogen Fertiliser Usage

A dramatic increase has occurred in the application of nitrogen
fertiliser between 1974 and 1983. In 1986 it was estimated that
1.43 million tonnes of nitrogen were applied in England and Wales
and just over 50 per cent was applied in the eastern part of the
country.

At present the efficient farmers already apply the optimum level
of nitrogen to crops and grass, whilst a major breakthrough in
plant breeding techniques is not expected before the end of the
century. Consequently, apart from the less efficient farmers
increasing their fertiliser application rates towards optimum
levels, little change in the overall application of nitrogen
fertilisers is expected.

c)   Pesticide Usage

Little regional data on pesticide usage exists. However between
1971 and 1983 it is estimated that a general increase in pesticide
usage has occurred particularly fungicides and herbicides.

As farming profits fall there will be a need to reduce costs. It
is therefore likely that in future there will be a downward trend
in the quantity of pesticides applied per unit area of crops and
grass, even in those areas where intensive agriculture remains.

d)   Genetic Engineering and Biological Control Systems

Informed opinion believes these new systems of production are
unlikely to be introduced for 10 to 20 years and extensive use is
unlikely to occur until about 2020. The introduction is likely to
reduce the need for inorganic nitrogen and pesticides and will

also have a major impact on land use throughout the country. However, at this stage it is impossible to predict what further impact these may have on land use in England and Wales.

6.2.2   Trends in Pollution from Agricultural Sources

a)   Nitrate

Because trends in land use change are at present difficult to predict, and because the relationship between land use change and nitrate levels have yet to be fully determined, likely changes in nitrate levels in groundwater cannot be predicted with accuracy. Also, nitrate investigations have not defined exactly the time-dependent aspects of trends in nitrate concentrations though an increase in nitrate levels over the past 30 years coincides with increased use of agricultural fertilizers.

The data shows that the position in different parts of the country is variable with examples being cited of nitrate levels rising, stabilising or in some cases falling. The general picture however is that in many areas of heavy groundwater utilisation nitrate concentrations are rising to levels approaching or above currently acceptable levels. Though it is possible that land use changes may affect this trend beneficially, the effects are long term and the rise in nitrate levels is therefore expected to continue in many areas for the foreseeable future.

b)   Pesticides

There is insufficient data presently available to establish trends in the concentration of pesticides in groundwater. The present position is one of increased incidence of occurrence of pesticides

in a situation where sampling and analysis techniques may not be
fully reliable.

It is increasingly apparent however as more sampling and analyses
are carried out that the problem is very widespread, and is likely
to pose significant difficulties to the water authorities in terms
of their ability to meet water quality standards.

c)   Other Pollution from Agricultural Sources

As discussed in Section 3.4 point source pollution from
agriculture is a relatively minor problem and although incidents
will inevitably occur there is no reason to suppose that the
situation will deteriorate.

6.2.3   Landfill

Landfill sites are a major threat to groundwater quality and a
number of cases of pollution attributable to landfill are
recorded. Given the prevalence of landfill as a method of waste
disposal it is surprising that the incidence of pollution problems
is not significantly greater. We consider that this must in part
be due to appropriate legislation and the authorities responsible
for its implementation.

The pressure on authorities to find sites for waste disposal is
however a continual problem. There are many areas in the south of
England for example where it is difficult to find sites of low
vulnerability to groundwater pollution while being otherwise
environmentally acceptable. There is no sign of this requirement
diminishing, though developments in landfill technology may reduce
the environmental stress.

The threat to groundwater quality from landfill sites is therefore likely to continue for the foreseeable future.

6.2.4   Sewage Effluent and Sludge Disposal

Disposal of treated effluent and sludge does not appear to present a significant threat to groundwater quality at present. The principal problems are likely to occur with untreated effluent from septic tanks, the incidence of which is widespread and increasing. The greatest threat may be to private borehole supplies on the same or adjacent properties to the septic tanks though there could be examples of threats to public supply sources.

The threat to private or small community sources caused by septic tanks seems likely therefore to increase, but the risk to major public supply sources is likely to remain relatively insignificant.

The use of sewage sludge in agriculture is increasing and may pose a threat to groundwater in the future. The impending national code of practice for agriculture use of sludge must address this issue, as described in 4.2.4.

6.2.5   Accidental Forms of Contamination

This category covers all forms of inadvertent pollution of groundwater through leaks, spills, pipeline fractures, drainage and contaminated land.

The number of accidents, spillages careless disposal and leaks etc resulting in groundwater pollution continues to rise, with locally serious consequences. The effects of any one incident may have long term consequences and indeed, many recent discoveries of contamination by organic contaminants, relate to incidents which occurred many years ago. Soakaway drainage systems on the modern road network can only exacerbate the problem.

Pollution due to accidental or inadvertent causes is therefore likely to become an increasing problem and is identified as a major threat to groundwater quality.

6.2.6   Other Forms of Pollution

a)   Mine Drainage

The present situation regarding pollution threats from mine drainage and disposal of spoil is not well documented. Although problems do occur the main threat is to surface water and thus to groundwater mainly by surface water – groundwater interaction. Few direct groundwater problems are reported.

With the decline in coal and non-aggregate minerals mining in England and Wales and the control of mine discharges having been brought into COPA legislation, there is no indication that the situation is due to worsen.

b)   Saline Intrusion

As reported in Section 3.4.10, saline intrusion either from the sea or from lateral movement of old saline groundwater is under control. No additional problems are therefore envisaged.

c) Acid Rain

At present problems with acid rain are mainly localised and of limited significance. Although the effects of acid rain on groundwater may increase it does not at present appear to be of immediate significance in regional groundwater quality terms.

6.2.7   Overall Groundwater Quality

Groundwater quality in England and Wales is increasingly under threat as a source of inexpensive high quality water. At present, groundwater generally requires minimal treatment for it to be delivered to consumers at a level of quality which accords with water quality standards currently in force. However there is considerable evidence that this situation is unlikely to continue. The principal problems are likely to be:

o   nitrate; upward trends in many areas, frequently approaching current EC limits

o   landfill; increasing pressure for environmentally acceptable sites

o   organics; includes solvents, pesticides and other compounds the discovery of which in groundwater is increasing significantly.

In addition to actual changes in quality the trend has been for MAC values set under EC directives to be lowered, particularly the trace organic constituents referred to above. Thus, the net effect is for it to become increasingly difficult for groundwater to meet water quality standards.

A feature of the anticipated problem is the difficulty and expense of treatment processes to remove the individual contaminants. It is likely that groundwater will no longer be the inexpensive option which it has been for many years.

6.2.8   Legislation and Policy

The most important effects are likely to result from changes in water quality standards set by the EC, and privatisation of the water industry. The implications of privatisation are discussed in Section 4.5.

7. CONCLUSIONS

7.1 Groundwater Use

(a) Approximately 30% of water used for public supply in England and Wales is obtained from groundwater. In addition, it is an important independent source of water to industry and agriculture.

(b) In many cases groundwater has a significant strategic importance, supplying centres of population which have no convenient alternatives.

(c) The total of licenced abstractions is such that additions are limited by resources availability; total abstraction has therefore probably stabilised and is not likely to increase.

(d) Only changes in groundwater quality or water quality standards are likely to bring about a significant decrease in groundwater abstraction.

(e) In terms of use, of the total groundwater abstraction, the proportion used for public supply is increasing and, correspondingly, private industrial use has declined.

7.2 Groundwater Quality Monitoring and Management

(a) Nationally, water quality monitoring is disproportionately orientated towards surface water considerations and the special needs of groundwater quality sampling are frequently not taken into account.

(b)   There is no direct requirement under existing legislation to
        routinely monitor untreated groundwater quality and
        monitoring practice is very variable.

(c)   There is no national system for the archiving of groundwater
        quality data and each water authority currently operates an
        independent system.

(d)   Nationally, monitoring and investigation are insufficient to
        enable an understanding of the distribution and behaviour of
        many important contaminants.

(e)   Monitoring increasingly involves detection of substances for
        which it is difficult and expensive to sample and analyse.

(f)   The excellent systems of monitoring and management of
        groundwater quality through aquifer protection policies
        operated by some water authorities are a product of
        individual initiative as distinct from national policy.

(g)   No general conclusions on costs of monitoring can be
        conveniently derived because of the difficulty of separating
        costs of groundwater from total water quality monitoring.

7.3   Natural Hydrochemical Conditions

(a)   The natural hydrochemical conditions in British aquifers are
        reasonably well understood

(b)   The susceptibility of different aquifers to pollution can be
        defined; the most sensitive aquifers are those characterised
        by fissure flow systems and low buffering capacity.

7.4 Sources of Contamination

(a) The principal threat to groundwater quality from agriculture is from diffuse sources, that is, intensive agricultural practices which involve high applications of fertilizer and pesticides; point-source pollution is of minor importance.

(b) Unless agricultural policies change, little overall change in application of nitrogen fertilizers is anticipated and the level of nitrate in groundwater is expected to continue to rise.

(c) The extent of pesticide contamination has yet to be fully determined but appears to be a serious problem. Although a reduction in the agricultural use of pesticides is anticipated, the future trend in use of non-agricutural pesticides is uncertain. In any case, the persistence of some substances is such that the problem is likely to continue.

(d) Landfill sites continue to be a major potential threat to groundwater quality.

(e) The disposal of sewage effluent is not a major threat other than septic tanks which may cause local pollution problems.

(f) Groundwater pollution from accidents, leaks and other forms of inadvertant pollution are a serious problem as they frequently involve persistent organic contaminants; some of the recently discovered problems may relate to incidents occurring many years ago.

(g)   There is a lack of overall recognition of the threat to
        groundwater resources from accidents etc as typified by some
        road drainage systems.

(h)   Although a number of serious organic pollution incidents
        originating from defence establishments have occured, there
        is no indication that such places are a greater threat than
        private industry of a comparable size, storing and handling
        similar substances.

(i)   Discharge of minewater drainage on the coalfields frequently
        results in poor river water quality and has been known to
        adversely affect groundwater quality.

(j)   Saline intrusion is under control and no problems are
        anticipated.

(k)   Acid rain effects on groundwater do not at present pose a
        significant threat to groundwater quality.

(l)   Interaction of groundwater and surface water systems can
        sometimes result in contamination of groundwater due to
        activities such as waste disposal, minewater drainage, and
        sewage effluent discharge to rivers.

7.5   The Current State of Groundwater Quality

(a)   The quality of groundwater used for public supply is
        generally good, requiring minimal treatment to meet current
        EC standards.

(b)   However, groundwater pollution problems have been identified
        and some are likely to worsen; (c) - (e) below are
        examples.

(c)   The principal inorganic contamination problem is nitrate
        with rising levels being reported in many areas of Eastern
        and Southern England and the Midlands.

(d)   Pollution in a variety of forms orginates from landfill
        sites, which are also a major potential threat. However
        there are few examples of sources actually being lost
        through landfill pollution problems.

(e)   The extent of organic contamination present as solvents,
        oils, pesticides etc is not known with certainty but appears
        to be widespread and poses a significant threat to
        groundwater quality.

7.6   Legislation and Policy

(a)   Current legislation relating to the protection of
        groundwater quality has generally been beneficial.

(b)   A loophole in the Control of Pollution Act 1974 means that
        the reinstatement and aftercare of waste disposal sites must
        be covered through the planning permission procedure, rather
        than through the site licensing. Water authorities are not
        statutory consultees under the Town and Country Planning
        Act.

(c)   Disposal of sewage sludge to land is a potential source of
        groundwater pollution but no problems have yet been
        identified. The DoE are presently preparing national
        guidelines for the use of sludge in agriculture, which
        should address the threat that the practice poses to
        groundwater quality.

(d)   Groundwater generally meets standards laid down in the EC
        Drinking Water Quality Directive with few derogations.
        However, there is considerable doubt within the water
        industry as to whether many of the standards relating to
        trace organic constituents are realistic and justified on
        the basis of toxicological research.

(e)   A few water authorities operate comprehensive aquifer
        protection policies. Adoption of such policies is clearly
        advantageous and a prerequisite for effective groundwater
        resource management and conservation.

(f)   The creation of a National Rivers Authority within a
        privatised water industry offers excellent opportunity for
        conservation, protection, monitoring and overall management
        of groundwater resources to be coordinated nationally.

7.7   Research

(a)   The present system of funding does not permit a fully
        coordinated groundwater quality research programme.

(b)   Present research policy is subject to political changes and
        because of the long time span of many problems this is
        detrimental to research into groundwater quality.

(c)   There is no formal medium of exchange of research information.

(d)   Research organisations do not at present take full advantage of EC funding opportunities.

(e)   The hydrochemical characteristics of major groundwater sources are sufficiently well understood for problems of contamination to be identified.

(f)   Work on trace elements in groundwater is proceeding and is providing useful baseline values.

(g)   Nitrates research has been effective and is now extending to land vulnerability.

(h)   A very good understanding of modelling of nitrate transport has been achieved for distributed and site specific forecasting.

(i)   Research into organic concentrations has only recently commenced and in this respect Britain is lagging behind continental Europe and the USA. Progress in organic research needs to be accelerated.

(j)   A wealth of information on landfills exists which can be built on in order to meet the additional research needs identified.

(k)   Groundwater quality modelling techniques are sufficiently well developed within the context of data available on key parameters.

8. RECOMMENDATIONS

8.1 Introduction

The principal conclusions of this study are that groundwater is a national water resource of major importance which is threatened by deterioration in quality due to contamination. Measures are required to investigate, monitor, manage and protect groundwater to avoid its irreversable loss.

All the recommendations set out in Section 8.2 are considered to be priority issues.

8.2 Summary of Recommendations

8.2.1 General Management and Co-ordination Aspects

(a) The value of groundwater as a national resource can be demonstrated and its conservation is of critical importance. However, there is no formal medium of exchange of information concerning groundwater management and aquifer protection. The success of existing systems is attributable to the skills, experience, initiative and enthusiasm of individuals within the water authorities who have particular interest in such matters. There is at present no national system through which management and conservation of the nation's groundwater resources is coordinated, when one would be desirable (Reference Section 3.2).

(b) We therefore recommend establishment of a system through which national management of groundwater resources can be achieved, with the object of conserving this important resource. Further consideration is required as to how this

111

might best be achieved, and should involve consultation with resource managers within the water authorities. Full advantage should be taken of any opportunities created by formation of a National Rivers Authority to meet this recommendation (Reference section 4.5).

8.2.2   Aquifer Protection

(a)   An essential feature of national conservation is development of an Aquifer Protection Policy. The advantages of a comprehensive Aquifer Protection Policy based on sound hydrogeological principles are clearly demonstrable in those authorities where such a system currently operates. It is recommended that a national policy on aquifer protection be adopted (Reference Section 4.3).

(b)   It is emphasised that whole-aquifer protection is required in addition to stringent protection measures around or near abstraction wells (Reference Section 4.3).

(c)   Further consideration of how the policy may be formulated and implemented is required, based on a detailed critical review of existing policies and consultation with the water authorities using them.

(d)   As part of these developments preparation of maps delineating aquifers most susceptible to pollution (land vulnerability mapping) should be extended to cover all major aquifers.

8.2.3   Monitoring

(a)   Monitoring of groundwater quality requires substantial improvement and it is recommended that a national network of observation points is established. Such a monitoring network could be managed through the National Rivers Authority, who could also have the responsibility for collation and archiving of data (Reference Section 3.2).

(b)   Further consideration is required to establish details of the network including number of observation points, parameters which need to be measured, where analyses will be carried out etc.

(c)   If the NRA are involved, the sampling and analysis aspects of the monitoring are likely to be carried out following competitive tender. If so, it is imperative that the contract specifications are sufficiently detailed for the difficulties associated with groundwater as distinct for surface water (Reference Section 4.5.2).

8.2.4   Legislation

(a)   The framework for waste disposal site licensing under COPA I needs to be extended to include reinstatement and aftercare matters. The water authorities (later NRA) should be made statutory consultees with respect to monitoring and control of construction, operation, reinstatement and aftercare (Reference Section 4.2.4).

(b)   Other activities such as industrial development, roads drainage etc have been identified as threats to groundwater quality; consideration needs to be given as to how these threats might be identified at the planning stage and the water authorities (or NRA) consulted, without creating complex bureaucratic procedures (Reference Section 3.4).

(c)   Due recognition should be given to views within the water supply industry that the levels of some constituents specified in the EC Drinking Water Quality Directive are unrealistic in terms of their perceived effect on human health (Reference 7.6 (d)).

8.2.5   Research (Reference Chapter 5)

The various research areas discussed in Section 5 are summarised below with respect to short-term and long-term requirements. It is difficult to define the time necessary for research but short-term is taken as up to two years for definitive results while long-term is assumed to be strategic ongoing research but which nevertheless progressively produces results within a properly structured time frame.

Short Term Requirements

(a)   Establishment of scientific methodology procedures for landfill investigations.

(b)   Appraisal of the reliability of existing hydrochemical sampling systems.

(c)   Development of sampling techniques and equipment suitable for studies of organic contaminants.

(d) Development of cheap in-well multiple depth sampling configurations which are also capable of measuring groundwater pressures.

(e) Appraisal of pollution in private groundwater sources particularly of shallow aquifers.

Long Term Requirements

(f) Determination of hydrochemical processes controlling the migration of organic pollutants through the unsaturated and saturated zones; probably the most important long term requirement.

(g) Collation of base line data pertinent to groundwater contaminants, including brackish groundwaters.

(h) Assessment of the impact of land use changes and agricultural practice on nitrate contamination of groundwater.

(i) Optimisation of the interaction between agricultural practices, nitrate protection zones and denitrification treatment.

(j) Development of protection zone methodology for nitrates and eventually organic contaminants.

(k) Microbiological controls on contaminants which are at present very poorly understood but thought to have an effect.

(l) Renovation of polluted aquifers.

REFERENCES

REFERENCES

ANGLIAN WATER. 1985. Annual water quality report.

ANGLIAN WATER. 1986. Annual water quality report.

ANGLIAN WATER. 1987. List of sourceworks outputs for Cambridge, Norwich, Colchester, Oundle, Lincoln.

ANON. 1987. Major groundwater source in Oxfordshire. Br. Geologist, 13, 103-4.

ANON. 1987. Water authorities blame low fines for flood of pollution : review of WAA publication 'Water pollution from farm waste 1986 England and Wales'. Surveyor, 27 August, 2.

BAXTER, K.M. & CLARK, L. 1984. Effluent recharge : the effects of effluent recharge on groundwater quality. Water Research Centre technical report TR 199

BRANDON, W.(ed). 1986. Groundwater : occurrence, development and protection. Institution of Water Engineers and Scientists water practice manual, no. 5.

BRITISH GEOLOGICAL SURVEY. 1986. Hydrogeological map of South Wales, scale 1:125,000.

BUNCE, R.G.H.(ed). 1987. Ecological consequences of land use change : Advisory Group meeting, Department of Environment, 16 November.

CENTRAL WATER PLANNING UNIT. 1980. Re-use of potable water supplies : final report. Toxicological aspects of synthetic organic chemicals in drinking water supplies, Section 3.

COLNE VALLEY WATER COMPANY. 1987. Groundwater resources study (Thames Water) : final documentation for tenderers.

COMMISSION OF THE EUROPEAN COMMUNITIES. 1979. Council Directive on the protection of groundwater against pollution caused by certain dangerous substances (80/68/EEC)

COMMISSION OF THE EUROPEAN COMMUNITIES. 1982. Groundwater resources of the European Community : synthetical report. (EUR 7940 EN)

COMMISSION OF THE EUROPEAN COMMUNITIES. 1982. Groundwater resources of the United Kingdom. (EUR 7946 EN)

COMMISSION OF THE EUROPEAN COMMUNITIES. 1986. Council Directive on the protection of the environment, and in particular of the soil, when sewage sludge is used in agriculture (86/278/EEC).

DEPARTMENT OF AGRICULTURE & FISHERIES FOR SCOTLAND. 1985. Control of Pollution Act 1974 : Part II (Pollution of water) : code of good agricultural practice approved by the Secretary of State for Scotland for the purposes of Section 31(2)(c) of the Act.

DEPARTMENT OF THE ENVIRONMENT. 1974 - 1978. Water Data Unit: Water Data. (5 vols)

DEPARTMENT OF THE ENVIRONMENT. 1980. Pesticide wastes : a technical memorandum on arisings and disposal including a code of practice. Waste management paper no. 21

DEPARTMENT OF THE ENVIRONMENT. 1982. EC Directive relating to the quality of water intended for human consumption (80/778/EEC). Joint circular DoE (20/82) and Welsh Office (33/82).

DEPARTMENT OF THE ENVIRONMENT. 1982. EC Directive on the protection of groundwater against pollution caused by certain dangerous substances (80/68/EEC). Joint circular with DoE 4/82 and Welsh Office 7/82

DEPARTMENT OF THE ENVIRONMENT. 1983. Agriculture and pollution : Government response to the Seventh Report of the Royal Commission on Environmental Pollution. Pollution paper no. 21.

DEPARTMENT OF THE ENVIRONMENT. 1984. Results of the survey of land for mineral working in England, 1982.

DEPARTMENT OF THE ENVIRONMENT. 1984. EC Directive relating to the quality of water intended for human consumption (80/778/EEC). Joint Circular DoE (25/84) and Welsh Office (51/84)

DEPARTMENT OF THE ENVIRONMENT. 1985. Mineral workings : legal aspects relating to restoration of sites with a high water table. Joint Circular DoE (25/85) and Welsh Office (60/85).

DEPARTMENT OF THE ENVIRONMENT. 1985. Water and the environment. Joint Circular DoE (18/85) and Welsh Office (37/85).

DEPARTMENT OF THE ENVIRONMENT. 1986. Landfilling wastes : a technical memorandum on the legislation, assessment and design, development, operation, restoration and disposal of difficult wastes to landfill including the control of landfill gas, economics, a bibliography and glossary of terms. Waste management paper no. 26.

DEPARTMENT OF THE ENVIRONMENT. 1986. Nitrate in water : report by the Nitrate Co-ordination Group of the Central Directorate of Environmental Protection. Pollution Paper no. 26.

DEPARTMENT OF THE ENVIRONMENT. 1986. Privatisation of the water authorities in England and Wales : report by DoE, Welsh Office, and MAFF. Cmnd 9734.

DEPARTMENT OF THE ENVIRONMENT. 1986. Seminar on research into groundwater quality, 27 November.

DEPARTMENT OF THE ENVIRONMENT. 1986. The water environment : the next steps : the Government's consultative proposals for environmental protection under a privatised water industry. (DoE & Welsh Office).

DEPARTMENT OF THE ENVIRONMENT. 1986. Water research in the longer term : report of the Long Term Water Research Requirements Committee.

DEPARTMENT OF THE ENVIRONMENT. 1987. Geology and minerals planning research programme projects completed, in progress and to be let.

DEPARTMENT OF THE ENVIRONMENT. 1987. Landwaste research programme 1987/88 : landfill sub-programme : research funds.

DEPARTMENT OF THE ENVIRONMENT. 1987. National Rivers Authority : the Government's proposals for a public regulatory body in a privatised water industry : report by DoE, MAFF, and Welsh Office.

DEPARTMENT OF THE ENVIRONMENT. 1987. Water Directorate research : six DoE/BGS contracts : progress reports, April-September.

DEPARTMENT OF THE ENVIRONMENT : WATER DATA UNIT. Water data 1974, 1975, 1976, 1977, 1978 (5 volumes).

EMINTON, S. 1986. Nitrate pollution of groundwater : BGS forecasts continuing rise.
Wat. Bull., 5 September, 8-9.

FIELDING, M., et al. 1981. Organic micropollutants in drinking water.
Water Research Centre technical report TR 159.

FOLKARD, G.K., et al. 1984. Investigation into chlorinated hydrocarbon solvents in groundwaters : final report to DoE for Contract PECD 7/7/-088.
London, Imperial College.

FOSTER, S.S.D.,et al. 1986. Groundwater nitrate problem : a summary of research on the impact of agricultural land-use practices on groundwater quality between 1976 and 1985.
Hydrogeological report of the British Geological Survey, No. 86/2.

GARDINER, J. & MANCE, G. 1984. Proposed environmental quality standards for List II substances in water : introduction.
Water Research Centre technical report TR 206.

GARDINER, J. & MANCE, G. 1984. United Kingdom water quality standards arising from European Community directives.
Water Research Centre technical report TR 204.

SIR WM. HALCROW & PARTNERS LTD. & LAURENCE GOULD CONSULTANTS LTD. 1987. Assessment of groundwater quality in England and Wales : research contract PECD 7/7/227 : inception report.

HARRIS, R.C. 1986. Changing attitudes to leachate migration and water resource protection.
Severn-Trent Water Authority report 0213R.

HARRIS, R.C. & SKINNER, A.C. 1986. The statutory controls over waste disposal
    and the role of the engineering geologist.
    Severn-Trent Water Authority report 0223R.

HARRIS, R.C. 1987. Leachate migration and attenuation in the unsaturated
    zone of the Triassic sandstones.
    Severn-Trent Water Authority report 0452R.

HEADWORTH, H.G., et al. 1980. Contamination of a Chalk aquifer by mine
    drainage at Tilmanstone, East Kent, UK.
    Q. J. Eng. Geol., 13, 105-117.

HEADWORTH, H.G. 1986. Groundwater protection policies in European countries.
    European Institute for Water seminar on Groundwater protection, policy
    and management, Strasbourg, March.

HEADWORTH, H.G. 1986. The South Downs Chalk aquifer : its development and
    management.
    J. Instn. Wat. Engrs. & Scientists, 40, (4), 345-361.

HOUSE OF COMMONS. 1987. Effects of pesticides on human health : 2nd
    special report of the Agriculture Committee, Session 1986-87.

HOUSE OF COMMONS. 1987. Pollution of rivers and estuaries : 3rd report
    from the Environment Committee, Session 1986-87.

HOWARD HUMPHREYS & PARTNERS. 1987. Management of groundwater abstraction :
    final report to Yorkshire Water.

INESON, J. 1970. Development of ground water resources in England and Wales.
    J. Instn. Wat. Engrs., 24, (3), 155-177.

INTER-RESEARCH COUNCIL COMMITTEE ON POLLUTION RESEARCH. 1986. Pollution
    research and the Research Councils : 11th report.

JOHNSON, P. 1986. More about privatisation.
    Circulation, newsletter of BHS, August, 1-2.

JONES, G.P. 1971. Management of underground water resources.
    Q. J. Eng. Geol., 4, 317-328.

JONES, G.P. 1976. Utilization of ground water in Britain.
    Groundwater quality, measurement, prediction and protection :
    conference, Water Research Centre.

KENRICK, M.A.P., et al. 1985. Trace organics in British aquifers :
    a baseline survey.
    Water Research Centre technical report IR 223.

KINNIBURGH, D.G. & EDMUNDS, W.M. 1984. Susceptibility of UK groundwaters to
    acid deposition : report to DoE.
    Hydrogeological report of the British Geological Survey.

LAWRENCE, A.R. & FOSTER, S.S.D. 1987. Pollution threat from agricultural
     pesticides and industrial solvents : a comparative review in relation
     to British aquifers.
     Hydrogeological report of the British Geological Survey no. 87/2.

MINISTRY OF AGRICULTURE, FISHERIES AND FOOD. 1986. Pesticides : guide to
     the new controls (Control of Pesticides Regulations 1986).
     Leaflet UL79.

MONKHOUSE, R.A. & RICHARDS, S. 1979. List of aquifers in England and Wales.
     In CENTRAL WATER PLANNING UNIT. Groundwater resources of the UK,
     Table 20.2.

MOREL, E.H. & WRIGHT, C.E. 1978. Methods of estimating natural groundwater
     recharge.
     Central Water Planning Unit technical note no. 28.

MORGAN-JONES, M. & GRAY, E.M. 1980. Comparative study of nitrate levels at
     three adjacent ground-water sources in a chalk catchment area west of
     London.
     Ground Water, 18, (2), 159-167.

MORGAN-JONES, M. & EGGBORO, M.D. 1981. Hydrogeochemistry of the Jurassic
     limestones in Gloucestershire.
     Q. J. Eng. Geol., 14, 25-39.

MORGAN-JONES, M., et al. 1984. Hydrological effects of gravel winning
     in an area west of London.
     Ground Water, 22, (2), 154-161.

MORGAN-JONES, M. 1985. Hydrogeochemistry of the Lower Greensand aquifers
     south of London.
     Q. J. Eng. Geol., 18, 443-458.

NATIONAL WATER COUNCIL. 1982. Water industry review. Supporting analysis.

NORTH ATLANTIC TREATY ORGANISATION. 1987. Demonstration of remedial action
     technologies for contaminated land and groundwater : NATO/CCMS Pilot
     Study : 1st International Workshop, Karlsruhe, Federal Republic of
     Germany, 16-20 March.

NORTHUMBRIAN RIVER AUTHORITY. 1973. Report on survey of water resources :
     periodical survey to conform with the requirements of the Water Resources
     Act 1963, Section 14.

PACKHAM, R.F. 1985. Drinking water quality and its significance for health :
     remedies and removal.
     Water Research Centre special subject no. 11

PALMER, R.C. 1987. Groundwater vulnerability : Map 1, Stafford.
     Soil Survey of England and Wales.

RAE, G. 1978. Mine drainage from coalfields in England and Wales : a summary
     of its distribution and relationship to water resources.
     Central Water Planning Unit technical note 24.

ROYAL COMMISSION ON ENVIRONMENTAL POLLUTION. 1979. Agriculture and
    pollution : 7th report of the Commission; Chairman Sir H Kornberg.
    Cmnd 7644.

SAYERS, M.A. 1987. Groundwater quality monitoring network - 1987 review.
    Severn-Trent Water Authority report Q444R.

SAYERS, M.A. & SKINNER, A.C. 1987. Pollution of groundwater by organic
    solvents at Shelton : report of a modelling study.
    Severn-Trent Water Authority groundwater report no. 28.

SEVERN-TRENT WATER AUTHORITY. 1986. Aquifer protection policy (draft).
    Report Q228R.

SEVERN-TRENT WATER AUTHORITY. 1987. Groundwater quality in Severn-Trent.
    Report Q432R.

SEVERN-TRENT WATER AUTHORITY. 1987. Landfill and groundwater protection.
    Report 6929T.

SKINNER, A. 1987. Groundwater protection and agriculture in the United
    Kingdom.
    European Conference on Impact of Agriculture on Water Resources,
    Berlin, September.

SKINNER, A. 1987. Policy for aquifer protection.
    International Association of Hydrogeologists (Irish Group) 7th Annual
    Groundwater Seminar, Portlaoise, 6-7 April.

SOUTH WEST WATER AUTHORITY. 1978. Survey of existing water use and management
    under Section 24 of the Water Act 1973.

SOUTH WEST WATER AUTHORITY. 1983. The drinking water quality sampling
    programme.

SOUTH WEST WATER AUTHORITY. 1983. The groundwater protection policy.

SOUTHERN WATER AUTHORITY. 1980. Survey of existing water use and management,
    under Section 24(1a) of the Water Act 1973. 2 Parts.

SOUTHERN WATER AUTHORITY. 1985. Aquifer protection policy.

SOUTHERN WATER AUTHORITY. 1985. Report on Thanet nitrate investigation : a
    study of the occurrence and cause of high concentrations of nitrate in
    groundwaters on the Isle of Thanet and their future trends.

SOUTHERN WATER AUTHORITY. 1986. Standard statistics 1985.

STATUTES. 1963. Water resources act.

STATUTES. 1973. Water act.

STATUTES. 1974. Control of pollution act.

STATUTES. 1983. Water act.

STATUTORY INSTRUMENTS. 1977. The town and country planning general development order (SI 1977 no. 289), and amending orders (SI 1980 no. 1946, SI 1981 no. 245, SI 1981 no. 1569).

THAMES WATER AUTHORITY. 1987. Water quality monitoring statement.

UNITED NATIONS ECONOMIC AND SOCIAL COUNCIL. 1986. Ground water management : note prepared by the Secretariat for ECE Committee on Water Problems, 18th session, February 1987.
Report Water/R.142, restricted.

UNITED NATIONS ECONOMIC COMMISSION FOR EUROPE. 1986. Ground-water legislation in the ECE region : a report prepared under the auspices of the ECE Committee on Water Problems.
UN publication E.86.II.E.21.

WATER AUTHORITIES ASSOCIATION. 1985. Waterfacts (also subsequent volumes for 1986, 1987)

WATER COMPANIES' ASSOCIATION. 1987. Evidence to the Royal Commission on Environmental Pollution study of fresh water quality.

WATER RESOURCES BOARD. 1973. Water resources in England and Wales, volume 1, report.
WRB publication no. 22.

WELSH WATER AUTHORITY. 1979. Survey of groundwater.

WRIGHT, C.E. 1974. Combined use of surface and groundwater in the Ely, Ouse and Nar catchments.
Reading, Water Resources Board.

WRIGHT, C.E. (ed). 1980. Surface water and groundwater interaction : report prepared by the International Commission on Groundwater.
UNESCO studies and reports in hydrology no. 29.

YORKSHIRE WATER AUTHORITY. 1987. Analytical frequencies/parameters for routine analysis of YW groundwater sources.

YORKSHIRE WATER AUTHORITY. 1987. Aquifer protection policy (draft).
Report JB 005 231.

YORKSHIRE WATER AUTHORITY. 1987. Details of abstraction licences and estimates of actual abstraction, 1977-1986.

YORKSHIRE WATER AUTHORITY. 1987. Details of public supplies, 1977-1985.

APPENDIX A

TERMS OF REFERENCE

SCHEDULE 1

GROUNDWATER QUALITY STUDY

OBJECTIVES

1. To provide for England and Wales an overview statement on the current state of groundwater quality in aquifers currently exploited or those potentially of use for drinking water supplies.

2. To highlight present and potential problems for managing groundwater, taking into account existing policies and practices for their use, and of legislation for their protection and regulation.

3. To propose issues for further examination including research.

PROGRAMME OF WORK TO BE CARRIED OUT BY THE CONTRACTOR

1. The contractor will be expected to exercise judgement in assessing the priority to be given to the subjects listed in presenting his proposals in accordance with the objective. The list is not necessarily complete and other issues may be introduced if considered by the contractor to be relevant to the study. The emphasis of the study is on the quality of groundwater and quantity aspects should only be considered where relevant to quality issues. Some aspects may be included if only to be dismissed as of little concern.

    a. <u>Current state of quality</u> Briefly review the state and trends of water quality in existing and potential water supply aquifers based on information readily available.

    b. <u>Uses</u> Briefly review and summarise uses of groundwater and their extent eg for drinking water supply, industrial water supply, agriculture, river regulation, waste liquid disposal, brine extraction, mine dewatering and other uses for which quality considerations are relevant. Consider implications of future uses in these categories.

    c. <u>Contamination</u> Briefly review and summarise the extent and implications of contamination of aquifers by point and diffuse sources. Consideration should include the incidence of nitrate, hydrocarbons, organic chlorine compounds and pesticides, accidental pollution, saline intrusion, land-use effects, leaking pipelines and sewers, storage sites, leachates at waste disposal sites and implications of motorway drainage and from surface mineral and mine workings.

    d. <u>Monitoring</u> Briefly describe present practices and costs for groundwater quality monitoring. Comment on the effectiveness of monitoring and suggest changes if appropriate.

    e. <u>Future problems</u> Present scenarios for and potential extent of problems and solutions for future groundwater management to provide potable water. Account should be taken of existing and proposed policies and legislation affecting groundwater quality particularly existing statutory duties of water authorities and the British Geological Survey.

    f. <u>Research</u> Briefly review existing and recent research and indicate future local and national needs.

2. The Department will require 30 copies of the final report.

APPENDIX B

PERSONNEL ENGAGED ON THE STUDY

PERSONNEL ENGAGED ON THE STUDY

Personnel engaged on the study were as follows:

Sir William Halcrow and Partners Ltd

Project Director	D O Lloyd
	Partner
Project Manager	E Cooper
	Senior Hydrogeologist
Water Quality Specialist	Dr D W M Johnstone
	Director
Project Hydrogeologist	K J Harries
	Hydrogeologist

Laurence Gould Consultants

Agricultural and Land-Use Specialist	C H Mathias

Independent Consultants

Consulting Hydrogeologist (Research Specialist)	Prof J W Lloyd, Professor of Hydrogeology University of Birmingham

An organogram for the project is given in Figure 1 of this Appendix.

Figure 1

PROJECT TEAM

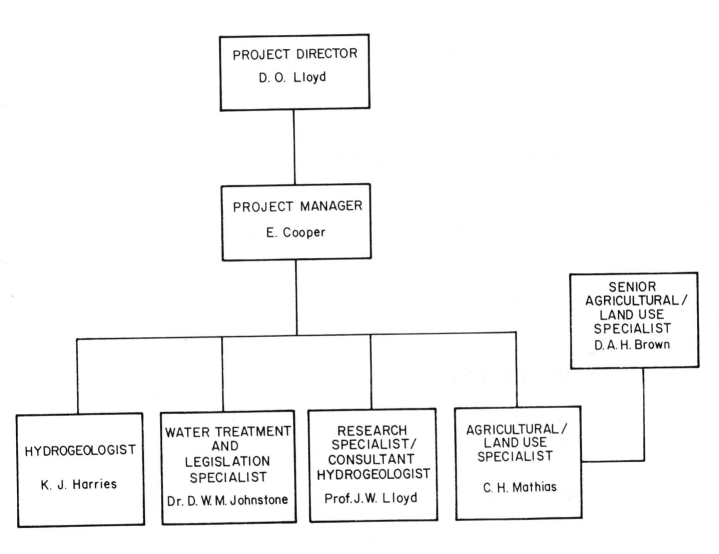

APPENDIX C

SOURCES OF GROUNDWATER CONTAMINATION

FROM AGRICULTURAL ACTIVITIES

APPENDIX C

SOURCES OF GROUNDWATER CONTAMINATION FROM AGRICULTURE

A   BACKGROUND

Since the war there has been a continuous increase in agricultural
productivity and production and as a result the United Kingdom is
now more than self sufficient in many temperate foodstuffs. This
improvement in productivity owes much to technological advances,
particularly over the last 10 to 15 years, due to the application
of research in such areas as improved crop varieties, more
effective use of fertilisers, better control of pests and
diseases, improved agricultural systems leading to a higher degree
of specialisation and greater mechanisation. In addition, better
education and training has led to improved management and hence a
faster uptake of new technology.

Agricultural activity is the source of various forms of
groundwater pollution. Unauthorised or accidental disposal of
animal and silage effluent, releases from chemical and pesticide
stores and disposal through soakaways have all occurred, sometimes
with serious local consequences. However, these incidents are
relatively minor compared with the impact of point source
pollution from agriculture on surface water. To date, agricultural
point source pollution has not caused permanent loss of
significant groundwater resources. However, diffuse pollution from
agriculture presents a far greater problem both in terms of its
current and likely future impact on groundwater quality.

B   NITRATE IN GROUNDWATER

1   Introduction

Research into the increase in nitrate concentrations observed in
groundwater from the major UK aquifers has implicated changes in
agriculture practices, including increased nitrogen fertiliser
application, as the prime source of nitrate. Other sources include
the atmosphere and sewage or industrial effluents, but their
contribution to nitrate in groundwater is insignificant in
comparison to that from agricultural practices.

2   Agricultural Practices Affecting the Level of Nitrate in
     Groundwater

The main agricultural practices affecting the level of nitrate in
groundwater are as follows:

a)   Fertiliser Application Rates and Cropping and Stocking
      Patterns

      The estimated leaching losses of inorganic nitrogen for
      selected winter and spring sown arable crops, together with
      grassland is presented in Table 1. This highlights the wide
      differential in the quantity of inorganic nitrogen leached
      from arable crops and grassland. The quantity of nitrogen
      leached is therefore highest in those areas where intensive
      arable cropping predominates.

TABLE 1 - ESTIMATED LEACHING LOSSES FROM SELECTED ARABLE

CROPS AND GRASSLAND

CROP	% NITROGEN LEACHED	ESTIMATED Kg/Ha of INORGANIC NITROGEN LEACHED (1983 APPLICATION RATES)
Winter Wheat	40	33
Winter Barley	40	20
Spring Wheat	50	71
Spring Barley	50	54
Maincrop Potatoes	50	102
Sugar Beet	50	77
Oilseed Rape	50	137
Cut Grass	10	16
Grazed Grass	15	19

Source: WRC/Survey of Fertilizer Practice/Consultants estimates
(Work undertaken by Rothamstead Experimental Station has indicated
that due to earlier sowing and a cessation of the practice of
applying autumn nitrogen, leaching rates from winter cereals
should be reduced by 40kg/ha).

The intensity of stocking on grassland also affects the level of leaching as the higher the number of livestock per unit area the greater the quantity of organic nitrogen produced. The nitrogen leached does not necessarily come directly from the nitrogen applied. Most of the fertiliser which is not taken up by the crop will initially be incorporated into the soil organic matter and then subsequently mineralised to leachable nitrate.

b)   Timing of Nitrogen Fertiliser Applications

The application of inoganic nitrogen, in appropriate quantities, applied in the early spring to early summer, when crop growth rates are high and the soil is below field capacity, is unlikely to cause direct leaching. Conversely, during the late summer and early autumn, soil nitrate supply usually exceeds the uptake by arable crops. Consequently, the application of nitrogen fertiliser to most arable crops between September and mid-February will contribute directly to the reservoir of nitrogen in the soil organic matter.

c)   Time of Planting

If no crops are planted in the autumn, all the soil nitrate released during this period will be leached. Thus the presence or otherwise of autumn sown crops has a major influence on the quantity of nitrate leached from arable land. Early planting in the autumn results in a greater uptake of nitrate by the crop.

d)   Ploughing up Grassland

The conversion of grassland to arable cropping is
significant in the nitrate balance due to the nitrate
released by ploughing. The oxidation of the soil organic
matter through ploughing releases large quantities of
nitrate which cannot be utilised by the succeeding crop and
is therefore leached from the soil profile. It is estimated
that ploughing up grassland releases the equivalent of 280
kg N per hectare over a period of some three years.

3   Climate

The concentration of nitrate in the leached soil water and
therefore the eventual concentration in groundwater depends not
only on the mass of nitrate lost from the soil but also on the
effective precipitation providing the recharge. Consequently, for
a given rate of leaching the nitrate levels in groundwater will be
higher in areas of low effective rainfall compared with the wetter
parts of the country due to the lower 'dilution' of nitrate in the
soil profile.

4   Trends in Land Use and Inorganic Nitrogen Fertiliser Usage
     in England and Wales

Trends in Land Use

Changes in land use in England and Wales over the last 20
years are summarised in Table 2. During this period the
total area of leys, permanent pasture and rough grazing has
declined by 14 per cent. In practice, most of the decline in
the area of rough grazing and permanent pasture will have
been due to their conversion to permanent pasture and

TABLE 2 – SUMMARY OF LAND USE IN ENGLAND AND WALES
– 1966 TO 1986 – '000 Ha

Year Land Use	1986	1976	1966
Cereals	3249.3	3155.4	3222.1
Oilseed Rape	276.3	47.8	-
Root Crops	339.9	380.0	375.7
Horticulture & Hops	199.0	278.0	253.9
Other Crops/Fallow	56.9	71.3	105.2
Crops for Stock Feed	231.0	216.8	232.1
Temporary Leys	1048.0	1369.1	1529.4
Sub Total Tillage	5580.4	5519.2	5718.4
Permanent Pasture	3870.4	4033.0	4126.2
Rough Grazing (sole rights)	1074.4	1185.7	1330.9
Woodland	228.4	167.9	N/A
Other Land	157.0	106.6	N/A
TOTAL	10910.6	11012.4	11175.5

N/A = Not Available
Source : MAFF June 4th Census

leys/incorporation into arable rotations respectively. Most of the reduction in grassland is represented by the fall in the area of leys, this area having been converted to arable cropping.

During the period under review the cereal area has shown an overall increase of 6 per cent. However, this increase has only occurred over the last decade as cereals have replaced the declining area of roots and horticultural crops. Oilseed rape was not grown 20 years ago but during the last decade the crop has increased by almost 500 per cent.

There is considerable variation in the distribution of crops and grass throughout England and Wales. A summary of the distribution of selected crops and grass in 1986 by standard statistical region is presented in Table 3. The boundaries of the standard statistical regions are shown in Figure 1. Table 3, and the summary Table 4, highlight the fact the arable cropping predominates in the drier eastern parts of the country (Yorkshire and Humberside to the South East) while grassland dominates in the west.

Figure 1

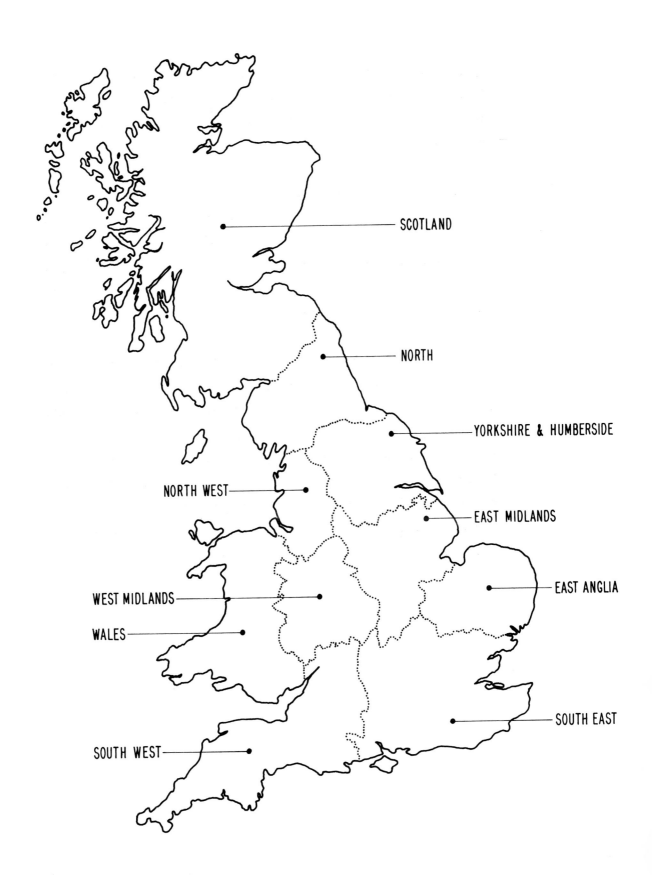

SCOTLAND

NORTH

YORKSHIRE & HUMBERSIDE

NORTH WEST

EAST MIDLANDS

WEST MIDLANDS

EAST ANGLIA

WALES

SOUTH EAST

SOUTH WEST

STANDARD STATISTICAL REGIONS

TABLE 3 - PER CENT OF CROPPING BY STANDARD STATISTICAL REGION IN ENGLAND AND WALES FOR SELECTED CROPS - 1986

CROP REGION	Winter Cereals	Spring/ Other Cereals	Oilseed Rape	Root Crops	Horticulture Hops	Leys	Permanent Pasture	Rough Grazing
	%	%	%	%	%	%	%	%
North	4	9	5	1	1	9	11	30
North West	1	4	1	3	5	5	6	5
Yorkshire & Humberside	13	12	17	14	9	7	8	13
East Midlands	19	11	27	21	21	7	7	4
East Anglia	17	14	14	40	25	2	2	2
West Midlands	9	9	7	11	8	12	10	2
South East	25	18	25	6	24	15	10	4
South West	11	17	4	3	6	27	23	9
Wales	1	6	-	1	1	16	23	31
Total Area - '000 Ha	2800.9	628.4	276.3	339.0	199.0	1048.0	3870.4	1074.4

1) Potatoes and Sugar Beet
2) Sole Rights

Source: MAFF June 4th Census

TABLE 4 - SUMMARY OF CROP DISTRIBUTION IN ENGLAND AND WALES - 1986

	East % of Cropping	West % of Cropping
Cereals	70	30
Oilseed Rape	83	17
Root Crops	81	19
Horticulture	79	21
Grass	27	73

These data emphasised the fact that current and potential nitrate problems will be greater in the east than the west due to:

- higher leaching of nitrate from arable crops;
- the reduced 'dilution' of nitrate in the soil profile in the drier eastern parts of the country.

b   Trends in Inorganic Nitrogen Fertiliser Usage

Apart from changes in land use there have been significant technological improvements in the agricultural industry, particularly over the last 10 to 15 years. As an example, the overall application of inorganic nitrogen fertiliser to arable crops and grass between 1974 and 1983 is presented in Table 5.

Table 5 shows that in 1974 the overall application of inorganic nitrogen fertiliser to both arable crops and grass was similar. However, since 1974 there has been a differential increase in the quantity of nitrogen applied to crops and grass. During this period the application of inorganic nitrogen per hectare of arable crops and grassland rose by 81 and 33 per cent respectively. This represents an overall increase on all crops and grass of some 53 per cent.

An estimate of the total quantity of inorganic nitrogen fertiliser applied in each standard statistical region has been undertaken for the years 1976 and 1986 and the results are presented in Table 6. The table shows that although there has been a considerable increase in the total tonnage of inorganic nitrogen applied over the last decade there has been almost no change in the percentage of inorganic

TABLE 5 - FERTILISER USE ON ARABLE CROPS AND GRASS - 1974 to 1983 - KG PER HECTARE OF PLANT NUTRIENT

YEAR	1974	1975	1976	1977	1978	1979	1980	1981	1982	1983
ARABLE CROPS	85	86	96	99	104	112	121	135	141	154
GRASS 1)	94	99	98	110	113	117	119	125	123	125
ALL CROPS & GRASS 1)	91	94	97	105	107	114	120	130	132	139

1) Excludes rough grazing

Source: Survey of Fertiliser Practice

TABLE - 6 ESTIMATED APPLICATION OF INORGANIC NITROGEN FERTILIZER

BY STANDARD STATISTICAL REGION IN ENGLAND AND WALES

- '000 TONNES PLANT NUTRIENT AND PERCENT

Year	1986		1976	
Region	'000 Tonnes	%	'000 Tonnes	%
North	107.6	7	68.8	7
North West	58.5	4	38.5	4
Yorkshire & Humberside	141.7	10	93.0	10
East Midlands	177.0	12	116.1	13
East Anglia	143-5	10	93.4	10
West Midlands	139.7	10	91.4	10
South East	237.8	17	158.3	17
South West	252.8	18	164.5	18
Wales	168.4	12	102.6	11

Source : Consultants estimates

A breakdown of the percentage of nitrogen fertiliser applied in 1986 between the east and west of England and Wales is given below.

	East	West
	%	%
Percent of Inorganic Nitrogen Applied	49	51

This shows that there is an almost equal split between the east and west in the total quantity of inorganic nitrogen applied. However, the estimates have been based on the assumption that uniform levels of nitrogen fertiliser are applied to crops and grass throughout the country. In practice it is likely that slightly higher levels of nitrogen are applied to arable crops by farmers in the east compared to their counterparts in the west. Consequently, it is estimated that over half of the inorganic nitrogen fertiliser is applied in the eastern counties of England and Wales.

C   PESTICIDES

1   Introduction

The other area of increasing concern from diffuse agricultural
pollution is the use of pesticides. At present, the
extent of pollution of groundwater in England and Wales by these
substances is not known but of late certain pesticide residues
have been detected in some aquifers, particularly in eastern
England. In this region, several groundwater supplies have been
found to exceed the EC MAC of total pesticides in drinking water
due to the presence of certain pesticides mostly of the
phenoxyalkanoic compounds (ie Mecoprop, MCPA, MCPB, and 2,4-D),
carbamates (Mancozeb and Mameb) and basic triazine groups. These
pesticides are mainly applied to cereals, potatoes and maize/beans
respectively.

To date, water undertakers have not attempted to monitor those
pesticides which might be present in groundwater supplies at very
low concentrations because reliable and sufficiently sensitive
methods of analysis are not available.

In July 1985 an EC Directive came into operation which set the MAC
for some 60 parameters relating to drinking water. As far as
pesticides and related products are concerned, the MAC for each
separate product is 0.1 micrograms per litre. The Government have
concluded that the MAC's for total and individual pesticides are
inappropriate because:

-    current analytical methods are unable to detect many
     pesticides even at concentrations above the MAC. Therefore,
     it is not possible to assess whether the MAC for total or
     individual pesticides has been exceeded;

-    no account is taken in the EC Directive of the widely differing toxicities of individual pesticides and it applies the same value to each pesticide irrespective of its toxicity.

To date the advice from the DHSS has indicated that, for those pesticides most frequently detected in water supplies and those most widely used, the concentrations being detected were several orders of magnitude lower than the Acceptable Daily Intake (ADI).

2   Farming Practices and Potential Leaching of Pesticides

In general, farmers are likely to adhere closely to the recommended application rates of pesticides for the following reasons:

-    in many cases there is a fairly narrow application 'window' at which the pesticide is effective. If the dose is too low the required effect is only partially achieved while if the application rate is too high the target crop can itself be damaged;

-    pesticides are expensive and farmers will tend to keep to the recommended application rates for this reason alone.

Thus the current legislation (the Food and Environment Protection Act 1985) mostly formalises the practices already undertaken by responsible farmers. However, scope exists to reduce the overall level of pesticide applications by persuading farmers to adopt a managed pest control approach, which has an element of risk, rather than a prophylactic approach which many farmers currently undertake.

The potential for leaching of pesticides and the possible risks
incurred will depend on the following factors:

-   the proportion of the spray which settles on the ground as
    compared with the target crop;

-   the potential for the pesticide to volatalise;

-   the composition of the soil profile and sub-soil;

-   the degradation products and whether these recombine to form
    other chemicals;

-   the biological activity of the degradation products.

Leaching of pesticides is therefore a very complex process and as
indicated above it is extremely difficult to identify the presence
of pesticides in groundwater due to the low levels of
concentration, lack of knowledge of the degradation products and
the fact that suitable analytical techniques are not currently
available. Although little data is available, it is thought that
the rate of leaching through the soil, sub-soil and aquifer
profiles is likely to be similar to that of nitrate.

3   Trends in the Application of Pesticides

Little regional data exists on trends in the application of
pesticides. However, since 1965, MAFF has undertaken a survey of
pesticide usage in England and Wales based on a sample of farms.
This data has been used to estimate the total usage of
pesticides, by type, in the two countries.

Specific data on pesticide usage is particularly difficult to summarise because:

-   although the national proportion of arable to grassland has altered relatively little in the last 10-15 years, changes have occurred within the area of arable crops grown, ie the swing from spring to winter cereals, the rapid expansion of oilseed rape and a decline in root crops;

-   changes have occurred in the formulation of agrochemicals;

-   there has been a rapid growth in alternative pesticides available to control a particular weed or disease.

As indicated above, farmers generally adhere closely to the manufacturers' recommended application rates per hectare. Variations in the total quantity of pesticides applied is therefore largely a reflection of changes within the arable sector, the uptake, by farmers, of new technology and the increased range of agrochemicals available to them. A summary of the estimated annual usage of pesticides on agricultural and horticultural crops, between 1971 and 1983, is presented in Table 7. An overall commentary is given below.

Insecticides, Acaricides and Molluscicides

The quantity of organochlorines used in 1980 to 1983 fell compared with the 1975 to 1979 period, although the area treated increased. This was mainly due to the usage of Gamma HCH on the expanding area of oilseed rape which masked a fall in the usage of other organochlorines. The quantity of organophosphates rose by 24 per cent between 1971/1974 and 1975/1979 but has remained relatively constant over the succeeding period.

TABLE 7 - ESTIMATED ANNUAL USAGE OF PESTICIDES ON AGRICULTURAL AND HORTICULTURAL CROPS IN ENGLAND AND WALES

PERIOD	1971 - 1974		1975 - 1979		1980 - 1983	
TYPE OF PESTICIDE	Treated Area 1) '000 ha	Tonnes Active Ingredient Applied	Treated Area '000 ha	Tonnes Active Ingredient Applied	Treated Area '000 ha	Tonnes Active Ingredient Applied
Organochlorine Insecticides, Acaricides	148	131	146	166	160	130
Organophosphate Insecticides	846	430	975	534	923	591
Pyrethroids	1	<1	41	2	172	8
Other Insecticides, Acaricides, Molluscicides	92	1286	556	905	956	664
Seed Treatments	3718	565	3753	591	3883	301
Fungicides	1895	2400	2253	2336	6715	4341
Herbicides, Defoliants 2)	6003	15250	7868	19925	12402	26360
Other Pesticides 3)	81	200	203	1038	801	3138
Area of Crops and Grass 4)	5631		9322		10511	

1) Total area treated, ie, if 10 ha has 3 applications, treated area = 30 ha
2) Includes chemicals for burning off crops
3) Growth regulators, soil sterilants, fumigants
4) Permanent pasture was not included in the 1971-1974 surveys

Source: Review of Usage of Pesticides in Agriculture and Horticulture in England and Wales (MAFF)

The fall in the tonnage of other insecticides, acaricides and molluscicides arose from the continued decline in the usage of tar oils in orchards and the replacement of many insecticides by pyrethroids which are applied at lower rates of active ingredient per hectare. However, within this group the use of carbamates has increased dramatically due to the use of methiocarb to control slugs in cereals.

Seed Treatments

Since the survey started, the use of chemicals for seed treatments has continued to show a slight increase. The fall in quantities used since 1979 is due to the replacement of ethirimol as cereal seed treatment with other systemic fungicides applied at much lower rates of active ingredients.

Fungicides

The application of fungicides to cereals came into prominence in the early 1970's but the overall levels of active ingredient applied changed little in the latter half of the decade. Since then, there has been a large increase in usage, particularly of foliar applied systematic fungicides to cereals. The 1980/1983 period also marked the widescale use of formulations containing systemic and dithiocarbamate fungicides which are frequently used against downy mildews.

Herbicides

During the period under review, large quantities of herbicides have been used in England and Wales and between 1971/1974 and 1980/1983 the tonnage of active ingredient applied has increased by 73 per cent. Most of this increase was on cereals and other arable crops although the greater area of grass included in the 1980/83 survey also meant that there was a larger treated area of grass. As the kg of active ingredient per treated hectare has not risen, the increase in total active ingredients applied is entirely due to the larger treated area of crops and grass.

Other Pesticides

Other pesticides such as growth regulators, soil sterilants and fumigants have shown a variable but upward trend in usage. Most of the recent increase is due to the use of chlormequat as a growth regulator in wheat.

4. Regional Distribution of Pesticide Usage

Little regional data exists on trends in the application of pesticides. However the following general comments apply:

a)   Pesticides are not normally applied to grassland except in the year in which the grass is established. Consequently, pesticides are predominantly applied in the eastern half of England. Although many farms in the Midlands and the west will have arable/livestock systems, the usage of pesticides per unit area of arable crops in these areas may well be below that in the east because:

-    many weeds found in predominantly arable rotations are not significant in arable/grass rotations;

-    in predominantly livestock areas, arable crops represent a relatively small part of the business and less intensive husbandry techniques are usually adopted compared with farmers in predominantly arable areas.

b)   As a result, trends in pesticide usage are dominated by the
agricultural techniques adopted by farmers in the major
arable areas in England.

5.   Other Users of Pesticides

Other organisations, particularly local authorities, public
utilities and nationalised industries also apply significant
quantities of pesticides. There is no recent collated information
on the non-agricultural uses of pesticides but it is believed that
the principal pesticides are creosote and tar oils for wood
preservation together with sodium chlorate and triazines. Within
England and Wales, the two latter groups of pesticides are
probably applied at a rate of several hundred tonnes of active
ingredient each per annum.

It is usually difficult to identify the source of pesticides in
drinking water because these substances are widely used. However,
the water authorities have indicated that the group of pesticides
found most regularly and at the highest concentrations are the
triazines. The use of these pesticides in agriculture is
relatively limited due to the fairly low area of crops to which
triazines can be applied. There is increasing evidence that the
occurrence of triazines in drinking water is largely due to non-
agricultural applications. The main use of these pesticides is
believed to be non-selective control of weeds on road verges,
playing fields, around housing and industrial estates and on
railway tracks.

D.   SEASONALITY OF NITRATE AND PESTICIDE CONCENTRATIONS

Seasonal variations in nitrate and pesticide concentrations mainly affect surface water quality. Although the concentrations of nitrate and pesticides will also vary over aquifers the, seasonal variations are more likely to be evened-out as the pollutants are transported through the soil, sub-soil and unsaturated zones of aquifers. However, for those aquifers where there is a significant recharge from surface water, seasonal variations in surface water quality will have an important impact on groundwater quality.

1.   Nitrate

During the late summer and early autumn, soil nitrate supply usually exceeds the uptake by arable crops; therefore the potential nitrate leached is at its highest during this period. Ploughing up grassland, particularly permanent pasture, also causes a sudden release of nitrate which is likely to continue for up to three years after the event. A further cause of an increase in nitrate pollution is the climate. Nitrate pollution can reach very high levels after a prolonged dry period. Drought conditions prevent crops taking up nitrate and the latter is 'flushed' out of the soil by the ensuing rain.

2.   Pesticides

Agricultural usage of pesticides is seasonal, and is usually confined to short periods of the year. In general, most herbicides are applied in the spring followed by fungicides and insecticides in the summer. Residual herbicides are

applied to winter cereals in the autumn. Seasonal factors
are not so important in the case of non-agricultural
pesticides usage.

3.   Monitoring

Analytical methods for certain well established pesticides
have been developed and number of surveys have been
conducted by water undertakers. However, unless monitoring
is carried out on a routine basis the seasonal peaks of
nitrate and pesticide pollution are likely to be missed.

E.   POINT SOURCE POLLUTION

As indicated earlier, point source pollution from agriculture is
relatively minor compared with the likely impact of diffuse
pollution. In addition, most point source pollution incidents
affect the quality of surface water although inevitably
groundwater contamination also occurs. The main contributors to
point source pollution are as follows:

1.   Pesticides

Pesticide pollution incidents are normally caused by
over-spraying water courses or drains, run-off from sprayed
or treated areas, careless disposal of empty containers or
washings from equipment, together with spillages either in
the field or chemical stores.

2.   Sheep Dipping

The disposal of water used for sheep dipping is another
potential source of pollution. However, the controls which
are exercised over sheep dips take account of the risks that
might arise from disposal.

3.   Animal and Silage Effluents

The development of livestock production systems over the
last 20 years has led to a considerable increase in the
intensity of production. Consequently, a substantially
higher stocking density gives rise to more animal excreta
per unit area from both intensive livestock and grazing
livestock enterprises. Similarly, improvements in forage
conservation systems have led to a decline in hay making and
an increase in silage production. This has necessitated
considerable development in the disposal of animal and
silage effluents.

The potential risk of accidental release of silage and
slurry effluent will depend upon the following factors:

-    management ability of the farmers;

-    the type of storage capacity of the system;

-    topography and soil;

-    the age and general standard of repair of the storage
     system;

- the distance of the structure from the nearest water course;

- drainage systems in the vicinity of the storage area;

- the possibility of exceptional rainfall events;

- weather conditions when slurry or farm yard manure (FYM) is spread.

4.   Sludge Disposal

The disposal of sewage sludge to farm land may form a potential pollution risk. However, there are guidelines for the disposal of sludge and to date there have been no adverse affects on groundwater.

F.   FUTURE TRENDS

1.   Introduction

Some groundwater sources already exceed the EC's nitrate MAC and others, although currently below the MAC, will exceed it at some point in the future due to the mass of nitrate already leaching through the subsoil and unsaturated zones. Any future change in land use or nitrogen fertiliser usage will therefore have little impact on the ultimate nitrate concentrations in these aquifers in the forseeable future. Conversely, the level of nitrate in other aquifers, particularly in the west, is relatively low but rising and future land use and fertiliser usage could well effect the ultimate levels of nitrate concentrations reached.

Future trends in pesticide levels in groundwater are difficult to predict due to the fact that only recently has it become apparent that there is a problem and current knowledge on movement of pesticides within aquifers and on degradation products is limited.

2.   Land Use

Future changes in land use will have a major impact on the distribution of the nitrate and pesticide pollution load. At present, a great deal of uncertainty faces the farming industry regarding the possibility of introducing either price reductions, increased quotas, set-aside etc or a combination of these. At present, no clear pattern is emerging regarding the likely outcome of negotiations between Member States of the EC or the wider international implications of the GATT negotiations.

Last spring the UK Government introduced the Alure programme which contains financial incentives to farmers who introduce on-farm woodland or diversify into non-agricultural activities. A recent proposal has been announced whereby payments may be made to farmers who take at least 20 per cent of their cereal land out of production. However, the introduction of the latter is dependent on the other EC Member States adopting a similar approach. At present it is extremely difficult to predict to what extent these financial incentives will be taken up by farmers or what geographical impact they will have on land use throughout England and Wales. In any event, economic and political influences are unlikely to be uniform throughout England and Wales and future land use change is best discussed under the following headings.

a)   Uplands

Ever since the lowlands were cleared, uplands have been
marginal for agricultural production. In addition, it has
been difficult to provide infrastructure and as a result,
these areas tend to have a high conservation value. Farmers'
ability to make a profit has for a long time therefore
depended on income support and in view of the conservation
value the public appears to be happy to regard farmers in
these areas as "custodians of the countryside" and are
therefore prepared to see them supported. Consequently apart
from further expansion of forestry there is likely to be
continued income support in these areas and any agriculture
land use changes will be relatively slow.

b)   Productive Lowlands

In England and Wales most of the productive lowlands are in
the former country. In these areas, there is little scope
for individual farmers to improve profitability by changing
their farm systems. Consequently, their best strategy is to
continue with their existing enterprises and where possible
achieve further technical and/or financial improvements.

Historically, technical performance has steadily improved.
This trend is likely to continue but at a reduced rate. If
profits are to be maintained, costs of production must fall;
therefore future technical improvements may be geared
towards higher production from lower inputs. In these areas,
farming will probably continue as the main rural activity
and changes in land use are likely to be relatively slow
unless extenisfication programmes such as set-aside are
introduced.

c)   Marginal Lowlands

Marginal lowlands will probably experience the most dramatic
change in land use when farm profits come under pressure as
price and other forms of support are reduced. These areas
are likely to include:

-       thin soils over chalk downland and other limestones;
-       light, droughty land without irrigation;
-       river valleys;
-       urban fringes;
-       heavy and wet lowlands;
-       sloping, arable land subject to severe risk of erosion.

The type of land use change that is likely to occur in these
areas will depend on location and in particular the
proximity of the area to a motorway. Urban fringes will
probably see the greatest changes as farming is less
attractive due to trespass, damage to property and
livestock, fragmentation and complaints from local
inhabitants.

In other marginal areas cereals are likely to return to
grassland, temporary grass will become permanent pasture and
existing permanent pasture revert to rough or extensive
grazing or even scrub. The area of woodland or forestry is
also likely to increase.

d)   Impact on Water Quality

A reduction in intensity of land use in river corridors and
on heavy wet lowlands is likely to improve the quality of
surface water. Groundwater quality may also benefit,

particularly if the aquifers are under thin soils on the
Chalk downland, limestones wolds or light sandy soils
without irrigation. However, the impact of a reduction in
farming intensity on groundwater quality may not be felt for
several decades.

3.   Inorganic Nitrogen Fertiliser Usage

The efficient farmers are already applying inorganic nitrogen
fertilisers at the optimum recommended rates. A major breakthrough
in plant breeding techniques is unlikely to occur before the end
of the century. Consequently, apart from the less efficient
farmers increasing their fertiliser applications towards the
optimum rates, relatively little change in the overall application
rate of inorganic nitrogen per unit area is expected. In those
areas where farming continues as the major rural industry,
cut-backs in nitrogen usage on arable crops are not considered
likely. There would have to be a dramatic reduction in
agricultural commodity prices before the current optimum nitrogen
application rates become uneconomic.

4.   Pesticide Usage

A reduction in pesticide usage is likely to occur even in those
areas that remain under intensive agriculture. At present, many
farmers apply pesticides on a prophylactic basis. As margins fall,
the need to reduce costs is likely to result in pesticides being
applied in response to a managed pest control programme.

5.   Genetic Engineering and Biological Control Systems

The major international agro-industrial companies are currently
investing heavily in genetic engineering and biological control

systems. Informed opinion believes that these new systems of production are unlikely to be introduced for 10 to 20 years and extensive use is unlikely to occur until about 2020. Clearly, the introduction of production techniques involving genetic engineering and biological control will have a major impact, not only on inorganic nitrogen and pesticide usage but also on potential land use throughout the country. Until the likely form of these new techniques is known it is impossible to predict what further impact these may have on land use in England and Wales.

Printed in the United Kingdom for Her Majesty's Stationery Office
Dd 289221   C10   11/88